"This book brings to us Taiichi Ohno's philosophy of workplace management—the thinking behind the Toyota Production System. I personally get a thrill down my spine to read these thoughts in Ohno's own words. My favorite part is his discussion of the *misconceptions hidden within common sense* and how management needs a *revolution of awareness.*"

Dr. Jeffrey Liker
Director
Japan Technology Management Program
University of Michigan
Author, *The Toyota Way*

"While no one person invented lean, no one is given more credit than Taiichi Ohno. Access to his true thoughts and ideas are rare, and this book is the best and most useful of Ohno's work. Many lean students would want nothing more than to spend a day with Taiichi Ohno walking through their plant. This book is the closest thing we have left to that experience. Jon Miller has done a diligent job, not just in translation, but ensuring that the true meaning comes through in a readable fashion. You truly feel as if you are in conversation with the father of the Toyota Production System. While this book won't paint a clear picture of what to do next on your lean journey, it should be required reading for any serious student of the subject."

Jamie Flinchbaugh
Co-author, *The Hitchhiker's Guide to Lean:
Lessons from the Road*

"This book and its translation provide the reader a wonderful opportunity to learn directly from the master architect of the Toyota Production System. One is able to hear, in his own words, the principles that have evolved into the most successful management method ever developed. Today, these lessons are being applied in many industries, including health care, in addition to their long-term application in manufacturing. This book enables the reader to get inside Taiichi Ohno's thinking as he makes concepts such as *kanban*, The Supermarket System, and Just in Time come alive in ways that can be easily understood. This book will help me, as a senior executive in health care, better implement our management method, the Virginia Mason Production System."

Gary S. Kaplan, MD
Chairman and CEO
Virginia Mason Medical Center

"Most of the chapters in *Workplace Management* can lead you to assume the 'revolution of awareness' Taiichi Ohno calls for is about lean specifics like customer focus, sensitivity to waste, increasing flow, and moving away from command-and-control management. But readers can see right in the first two chapters that Ohno is also suggesting we look back at ourselves and our mindset. Ohno espouses greater awareness not just about the lean goals we pursue but also about the habits and patterns of how we pursue them.

"Human capability for learning and change is astonishing, and I think Ohno was an optimist about that. But to mobilize that capability throughout an organization, and even society, we should acknowledge that our unconscious mindset and habits often drive us to try to solve problems in unscientific (overconfident, emotional, mechanistic) ways. When I read *Workplace Management* today that's as much a part of Ohno's message as the rest of the book, and I think the book endures, in part, because of that message."

Mike Rother
Author, *Toyota Kata* (McGraw-Hill)
Co-author, *Learning to See* (Lean Enterprise Institute)

Taiichi Ohno's Workplace Management

Special 100th Birthday Edition

With new commentary from global quality visionaries

New York Chicago San Francisco Lisbon London
Madrid Mexico City Milan New Delhi
San Juan Seoul Singapore Sydney Toronto

McGraw-Hill books are available at special quantity discounts to use as premiums and sales promotions, or for use in corporate training programs. To contact a representative, please e-mail us at bulksales@mcgraw-hill.com.

Taiichi Ohno's Workplace Management, Special 100th Birthday Edition

1 2 3 4 5 6 7 8 9 0   DOC/DOC   1 8 7 6 5 4 3 2

ISBN 978-0-07-180801-9
MHID   0-07-180801-9

This book is printed on acid-free paper.

Sponsoring Editor
Judy Bass

Copy Editor
Lisa McCoy

Editorial Supervisor
Stephen M. Smith

Proofreader
Marie Thompson

Production Supervisor
Richard C. Ruzycka

Indexer
Judy Davis

Acquisitions Coordinator
Bridget L. Thoreson

Art Director, Cover
Jeff Weeks

Project Manager
Patricia Wallenburg, TypeWriting

Composition
TypeWriting

CONTENTS

FOREWORD

Learn Ways of Looking at Things and Thinking about Things

While Taiichi Ohno is considered to have been an influential and revered business leader, to me he was a mentor as well as a strict teacher to be feared. One of my life's treasures is that I was able to learn the basic teachings of *genchi genbutsu* directly from him. The Toyota Production System pioneered by Mr. Ohno is not just a method of production; it is a different way of looking and thinking about things, and it has had a profound effect on my way of life.

Mr. Ohno was a thorough champion of workplace-led management and of following the principles of fact-based reason. Through persistent on-the-factory-floor trial and error, he built a system that relentlessly pursues the elimination of waste to realize cost reductions. His conviction was that the truth exists in the *gemba* (the workplace or where the action is happening), whereas theories are just products of imagination.

Mr. Ohno based his creation on this conviction and the belief that a company can't develop unless its people are nurtured. While he was in the process of creating the Toyota Production System, he gave management in the *gemba* (including me) *genchi genbutsu*–based practical tasks through which we were matched in a "competition of wits" against him. This is the hands-on human resources "nurturing" that he, a great educator, promoted.

The *genchi genbutsu* way of looking and thinking that Mr. Ohno taught me is told in this book. As I read each line, the memories come back and it's almost as if I am back there again. This book is Mr. Ohno's way of passing down all his knowledge and wisdom clearly and tenderly to future generations. It is truly a precious record.

From my personal experience of working in the United States and from my time managing a global company, I truly believe that the knowledge and wisdom contained within this book is valid not only for Japan, but holds true across all borders.

I sincerely hope that this book is helpful in human resources development and that it helps all who read it rethink how they look at and think about things.

Fujio Cho
Chairman
Toyota Motor Corporation

PREFACE

I was hired by Toyota Motor Corporation and became directly involved in the manufacture of automobiles 37 years ago in February 1945. When I think back to those days and the progress in automobile manufacturing since then, it seems like we are in a different world today. Having spent all of my time on the *gemba*[1] during those years, this progress seems normal to me. On the other hand, I think the progress has been immense.

However, when I think of 10 or 20 years into the future, the changes to come will be unimaginable to us today and there is no time to be sentimental. The past is the past and what is important is the current condition and what we will do next to go beyond where we are today. It is meaningless to compare before *kaizen*[2] and after *kaizen*.

By the way, it seems people refer to me as the founder of the Toyota Production System or the creator of the *kanban*[3] system. Indeed, we called it the Ohno System for a time when we were going through a period of trial and error to establish an innovative production system. However, the credit for the creation of the Toyota Production System rests with none other than Toyota Chairman Eiji Toyoda, the encouragements of the late Toyota Advisor Shoichi Saitoh,

---

[1] *Gemba* (pronounced with a hard "g" as in "go") literally means "actual place" and implies a location where the action happens or where value is created. *Gemba* is commonly translated as "shop floor" or "workplace." Ohno uses *gemba* to refer to the administrative workplace, the shop floor at Toyota, and also to the people who work on the shop floor.

[2] *Kaizen* is the Japanese word for "improvement" and in this context means continuous improvement.

[3] The *kanban* system is a material replenishment system based on a "pull" from the customer, rather than a "push" by the producer. *Kanban* literally means "sign board," for the cards that are used to signal the reorder of parts.

and the efforts of all of those people on the *gemba* who gritted their teeth at my complaining and gave their cooperation.

In a word, the Toyota Production System is to "produce what you need, only as much as you need, when you need." When you think about it, this is a very commonsense thing, but I think the fact that this is so difficult to do is because we are trapped by our habits and ways of doing things and we cannot change our ideas and our actions.

Although hardly deserving, in the spring of 1982 I was decorated with the Order of the Rising Sun, Third Class. On this and also on the occasion of having served for over 30 years at Toyota Motor Corporation, following the merger of Toyota Motor Company and Toyota Motor Sales, I have collected my experiences in this book in the hopes that the reader will find them useful. I am sure you will find awkward sentences as you read, but I hope you will gain some hints on how to break down your misconceptions.

Furthermore, this text was born from the strong urgings of Chairman Akira Totoki of the Japan Management Association and many others, and for this I would like to express my gratitude.

Taiichi Ohno
September 1982

CHAPTER 1

The Wise Mend
Their Ways

I have been asked to talk about the theme of "management of the *gemba*," and since I am not confident that I will be able to do this systematically, I will talk about related matters as they occur to me.

I don't think that the *gemba* changes easily. If the *gemba* changed easily, this would be very easy, but the *gemba* is not such a place. It is important for people to understand and agree, and it is important for us to persuade them.

In order to explain and gain the agreement of many people, you need to have some basis for your arguments. When I give talks I am often asked about how to develop one's powers of persuasion. But if you are in a position to give instructions or give orders, you cannot do this unless you have a lot of confidence about what you are saying.

However, people's ideas are unreliable things, and I would be impressed if we were right even half of the time.

There is a proverb: "Even a thief is right three times out of ten."

If it's true that even a thief will say three right things, then I think we should expect that a normal person is right half of the time but wrong the other half of the time.

When I was a middle school student in the old system,[4] we studied the Chinese classics, and during this class we learned from the *Analects of Confucius*. In these writings Confucius says, "The wise will mend their ways" and "The wise man should not hesitate to correct themselves." I think the term "wise man" must refer to a remarkable person, and I am sure that more than half of what such a person said was right, but even then they probably were wrong 30 percent or 40 percent of the time.

I think these words in the *Analects* mean that even the wise man is not right ten out of ten times, and when you know you are wrong you should mend your ways and not hesitate to correct yourself.

A thief may say good things three times out of ten; a regular person may get five things right and five things wrong. Even a wise man probably is right seven times out of ten, but must be wrong three times out of ten, so if you are wrong don't hesitate to correct yourself.

Confucius is saying that we should change gracefully, like a leopard. I think his words mean that in the end it is not good if you hold on to your ideas too strongly and try stubbornly to justify them.

There is another saying: "The morning's orders are revised in the afternoon." If my memory is correct, we were taught that it is a bad thing to give orders or instructions in the morning and then change them in the afternoon, but I think that as long as "the wise mend their ways" and "the wise man should not hesitate to correct himself," then we must understand this to mean that we should, in fact, revise the morning's orders in the afternoon.

However, this does not mean that you give ambiguous orders or instructions you are not confident about in the morning and then change these orders without even going to see the results. If you are giving orders, or if you have given an order, and you see by the result that you were wrong, or that circumstances have changed, making your orders bad, you should not wait until the afternoon to change these orders. Why not revise the morning's orders in the morning? What is necessary is the attitude that if the morning's orders were bad they should be changed by noon at the latest. From this point of view

---

[4] The old system middle school is equivalent to today's high school in Japan.

there are countries that pass regulations and laws and do not change them.

I encounter these from time to time and think, "I can't believe such laws still exist."

There may be many countries that believe that it is bad if "the morning's orders are revised in the afternoon" and leave the same laws on the books for many years, and I am sure this is true even locally. This does not mean that companies should also stick stubbornly to old ways and blindly follow established authority.

Engineers, in particular, tend to hold on tightly to things they have said or to their ideas. Engineers are often said to be inflexible or stubborn, but I think it is important for them to quickly correct themselves, just as the wise mend their ways. If you think, "What I said was mistaken," you should clearly say, "I was wrong." Without this sort of attitude your subordinates and the people on the *gemba* will not do things for you. If you realize that people will make mistakes and have a frank attitude to the point of thinking it is normal to apologize and say that you were wrong even to your subordinates, this will have an effect on how persuasive you can be.

If you fear the other person, or if you do not understand why, and you just keep going ahead, knowing you are wrong but doing nothing about it, you will not know what is really wrong. This has a negative influence over time. It becomes awkward to change the order you gave and so you leave it alone. As a result, people stop following you.

We are all human and we are wrong half of the time. You may give the wrong orders to your subordinates. Since we are all human, half of what your subordinates have to say may be right. Unless managers first take this attitude, people will turn away from us.

So in the end, having a sense of humility is one of the conditions for developing strong powers of persuasion.

CHAPTER 2

If You Are Wrong, Admit It

This raises the question of why we are wrong half of the time. This may be because even when we say something with a lot of confidence, many times our fundamental way of thinking is wrong.

In Japan, we have a word, *sakkaku*,[5] which is very appropriate. I think the optical illusion, or the misconception of what we can see, is easy to understand. For example, in the following diagram, if the two lines of equal length are made into a "T" everyone sees that the horizontal line looks shorter than the vertical line. This is a common method for making a most basic explanation of misconceptions. You can make mistakes when you think "this one is longer" because it looks longer.

However, we cannot help the fact that it looks longer to us. In these situations we have to take apart the "T" shape and arrange the two lines next to each other, and we will see that they are the same length. So, even though one line looks longer, in fact it is not longer.

The misconception of an optical illusion is very easy to explain, and people are easily persuaded.

---

[5] *Sakkaku* (錯覚) means "misconception."

The question then becomes, "How long should it look for the lines to be the same length?" or "How long should it look for the line to be longer?" This is not something that we can judge by sight alone, and again we need to arrange the two lines next to each other for comparison.

There are so many things in this world that we cannot know until we try something. Very often after we try we find that the results are completely the opposite of what we expected, and this is because having misconceptions is part of what it means to be human. While it is easy to persuade people by trying out the optical illusion, it is difficult to prove that the ideas in your mind and the thoughts in your brain are, in fact, misconceptions. In many cases when a person has an idea or makes a statement that they believe is correct, they find that it was a misconception. When you try your ideas the results can be contrary to your expectations.

As long as humans have their misconceptions, we are lucky if we give ten orders and half of them are correct. I think Confucius was able to say, "The wise should not hesitate to correct themselves" because he knew that we make mistakes half of the time.

People who hold the misconception in their head that one line is longer will not easily understand if you tell them the two lines are the same length. They just have to try it. Once they try it and verify the results with their own eyes, they will realize that the orders that they gave, believing they were right, were in fact wrong. They will also make the workers try many different things to help them understand misconceptions on their own.

When making people try things, it is important for the person who gave the instruction to go see the results with their own eyes. When verifying with your own eyes, if you see that it was not a misconception but was in fact true, and you can say, "I was wrong," on the spot, people will think, "He is my boss but he apologized to me when he was wrong."

As a result when you have another idea and you instruct them to try it out they will do so willingly.

If you are wrong and you show by your facial expression, "Well, I'll be danged," this will become a form of encouragement to them. As they try ten different things and they see that five of things you ask them to do are correct, I think they will become very cooperative.

On the other hand, if you insist stubbornly that the boss's orders should be followed, whether they are good orders or bad orders, people will stop following you. On the question of persuasion, when both the person giving the orders and the person being ordered recognize that as humans we are only right half of the time, we can say, "What did I tell you?" to the other when they wrong, and just this feeling of openness makes the person you are trying to persuade feel better. As a result, they will become more willing to cooperate. I think this is the true power of persuasion.

If people did not have misconceptions, there would be no need for persuasion. Because we fall into misconceptions due to ideas in our heads, persuasion can be difficult. Perhaps the more that a person is an intellectual the more they are prone to misconceptions.

CHAPTER 3

Misconceptions Reduce Efficiency

On the *gemba*, as I just mentioned, it is important to just try it. For instance, I am sure this is something you find everywhere, but people think that doing one type of work all at once is faster. When I tell people, "Do one piece at a time," they say that this will lower efficiency. They think that efficiency is improved—in other words, that productivity is improved—by producing the same thing over and over.

I was observing a young woman performing an inspection process, and she was arranging many parts in a row and checking them. No matter how much I told her that instead of doing it her way it was much easier and much more efficient to inspect and put them in a box one at a time she would say, "No, this way is faster."

In these situations, I say, "All right, that's okay, but try it my way one at a time." When people try it, they find it is too boring. They think maybe they will not be able to make their numbers this way. However, after they try this for a whole day, they find that what used to require overtime to complete 5,000 pieces can now be done one at a time every 20 seconds, in regular hours. It seems they cannot believe that they can get more done with such a slow pace of work.

When they work on many pieces at once, taking 20 or 30 pieces in one hand and arranging them neatly in rows, they have the misconception that they can get more work done. Again, we make them try working on one piece at a time. They may think, "This is not real work; this is play." As a result of working like it was play and finishing work in regular hours, the worker does not work overtime and their income is reduced. If their argument is that doing many at once is better because of their finances, I can't argue with that…. When working one piece at a time you can work at a leisurely pace that does not make you tired and you can make the same volume without overtime. They understand this when they try it. All of this is relatively simple. However, the reality is that this simple thing is not actually done on the *gemba*.

This is an old story from the Toyota Motor Company, right after World War II. At the process for drilling holes in round bar stock, the worker wanted to only drill holes. The daily requirement was 80 pieces, so the young worker was drilling the holes by manual feed. "Why is he operating the machine by manual feed?" I wondered.

The worker explained that on automatic feed the machine would keep going even after the cutting tool became dull and did not cut so well, and this caused the cutting tool to break or the dimension of the hole to be wrong. By operating the machine on manual feed the worker could tell how the tool was cutting. "So this way is faster," he said.

When I asked, "How long does it take to make the hole?" he replied "Thirty seconds." "So," I said, "if you can make a hole in 30 seconds, you can make two holes in one minute." The worker had nothing to say to this. The reason is that this job was done over seven hours of working time. The worker was proudly saying that he made 80 parts in seven hours. He was saying that he was operating the machine by hand and doing his best to make 80 parts in a day, as required.

Since there are 60 minutes in one hour, next I said, "You can make 120 holes in one hour." He did not reply because while he was proud of making 80 parts in a day, if it was possible to make 120 parts in one hour, this was troubling news to him. This is why he did not respond when I said, "You can make 120 holes in one hour." The message was, "Why do you only make 80 parts in seven hours? If you need 80 parts it should take you 40 minutes. This means you are only working 40 minutes in one day."

"I'm working diligently and doing what is needed. Why do you complain?" he asked me.

I said to him, "Son, you may be diligently working up a sweat but you are only making 80 parts in seven hours. If you are going to come to work, give us at least one hour of work per day."

"Give me a break," he said.

When you think about this, it may seem that making holes at the fastest speed you can by hand is the faster way. Making a hole using automatic feed takes 40 seconds. Making a hole by manual feed takes 30 seconds. So it seems that manual feed is more efficient. But after making three holes in a row, one after another, with the manual method, the tip of the drill gets hot, and this makes it dull. This causes it to cut not as well. The worker takes the cutting tool over to the grinder to sharpen the cutting tool, and then back to make three more holes. The cutting tool gets hot again after two or three holes, and the worker has to go back to the grinder to sharpen the tool. He thinks he is working.

He thinks that if he works diligently he can make one hole in 30 seconds. He has the misconception that doing the same task over and over again raises efficiency.

However, if you use the automatic feed and you only need 80 parts per day, you only need to make one hole every five or ten minutes.

The appropriate cutting speed is 40 seconds per hole. You can make a hole in 40 seconds and then let the cutting tool cool for four minutes and 20 seconds so that the cutting tool will return to room temperature when it is needed again to make the next part. You can also apply cutting fluid to the tool to lower the temperature to that of the coolant, and then you can use the same cutting tool to make 30 to 50 holes before needing to sharpen it again.

Each worker did not have their own whetstone. Five or six people lined up before the whetstone waiting to use it. Everyone was working in the same way. For example, the lathe was used to cut parts as fast as the cutting tool would allow. This dulled the cutting tool, and the lathe operators lined up to sharpen their tools. There were always five or six people in queue before the whetstone. So even if they could sharpen a tool in 30 seconds, if they were at the end of the queue of five or six people and each sharpened their tools, it took ten minutes before they got back to their machine. If a worker found that he did not sharpen the tool well and had to go back again, he might only make two parts in ten minutes.

The old drill presses had a small table, so if a worker tried to make many parts at once he would need to take 10 or 15 pieces of material out of the box to put them on the table of the drill press.

Then he moved 10 or 20 finished parts to the side and into a basket.

Again, he took 10 or 20 pieces from one basket and lined them up on the table. The person doing all of this thought he was working. As a result he could only make three or four pieces every ten minutes, and yet he thought that he was doing a good job because he could make a hole every 30 seconds, while it took 40 seconds on automatic feed. "And I am sharpening my own tools," he thought.

If a worker only needed one part every five minutes they could have let the cutting tool cool down for four minutes and only have gone to sharpen tools once per day, and yet they went to sharpen tools every three parts. Although they thought that they were skilled workers efficiently working up a sweat, in fact, this was a very inefficient way of working.

CHAPTER 4

Confirm Failures
with Your Own Eyes

It is relatively easy to persuade people on the *gemba* with examples like these, but away from the *gemba* there is not always a way to prove one's point, so many times each side ends up thinking their idea is a good one. Perhaps the hardest thing is for managers, senior managers, and supervisors to persuade each other.

For example, it can be difficult when an area manager must persuade a team leader, who is a factory floor supervisor, to try something. If the team leader is not persuaded, they will not instruct their workers to try it. Even if they think that the other person's idea has value, they cannot tell in their minds who is right and who has the misconception.

They become caught in endless debate, the *gemba* remains stuck in their old ways, and the productivity of that workplace does not improve. So just try it. Try it, and if there are two opinions, let them each try it their way for one day. Or try out the idea of a supervisor from another area.

To exaggerate a little, each person should be tenacious when testing their ideas and checking the results until everyone is persuaded that they have found the one better way.

This is not the same as stubbornly holding on to your ideas. If it is something you said or an idea you had, there is something good about it. You may have misconceptions, but you also have good ideas. If your idea fails, then go see what failed with your own eyes. It is important to develop this habit.

When managers only hear about the results and think, "Oh, that didn't work either. The old man doesn't know what he's talking about," then the result is failure. Even if the result is a success, you must not be satisfied by only hearing about the results. Go see with your own eyes, and you will understand very well what things were tried and what things were not included in your calculations.

After seeing that so much time was being spent on sharpening drill bits and cutting tools rather than on getting work done, we started thinking that we needed to set up a centralized grinding operation.

When we said we would set up a centralized grinding operation, one experienced worker said, "No, we tried that during the war, but it failed. That's why we do things the way we do now."

"I did not see it fail during the war. Show me again how it fails. If I am persuaded by this, I will let you continue doing it the way you do it now."

I said, "I think the reason it failed was because something went wrong. In those days our products were treated as military supplies.

"It failed because people from the army came and forced you to do centralized grinding, and you did it reluctantly. There is no way the results could have been good. Now I am asking you to do it, so let me see with my own eyes how it fails." We tried it, and it did not fail.

While we tried this, the experts sharpening the tools would try to tell me all sorts of things I did not know, such as how the number two tool must be sharpened for castings or what to do in the case of iron. But all of that is irrelevant to me, and has no effect on centralized grinding. The important thing is to make it clear to the person doing the grinding what the cutting tool is used for, at what machine and for what material it is used, and therefore what angle the cutting edge should be ground to and what material the cutting tools should be made of. These are documented as standards, and people need to follow the standards. If you think that each one of the hundreds of workers must be able to sharpen their own tools in order to be fully competent, this makes efficiency very poor.

We saw what caused failure, or what was about to cause failure, and took action ahead of time to prevent these things. As a result we avoided failure and multiplied production output by many times. The workers themselves could make good products without proprietary techniques or special skills.

This happened shortly after the war, so it is a very old story. I doubt that there are any workplaces today that do things that way.

Assuming that we each have misconceptions in our minds, it makes me think that the ability of people to relate to each other will become a significant strength.

CHAPTER 5

Misconceptions Hidden within Common Sense

These misconceptions easily turn into common sense. When that happens, the debate can become endless. Or, each side tries to be more outspoken than the other and things do not move ahead at all. That is why there was a time when I was constantly telling people to take a step outside of common sense and think by "going beyond common sense." Within common sense, there are things that we think are correct because of our misconceptions. Also, perhaps a big reason we do some of the general commonsense things we do is that based on long years of experience, we see there are no big advantages to doing things a certain way, but neither are there many disadvantages to it.

I like to think that if there are big advantages to something, there are also big disadvantages waiting for us to stumble upon them. If you can minimize disadvantages by being afraid of them, that would be all right. If we go beyond the common sense that when the advantage is small, it is all right as long as the disadvantage is small, we will see that where there is a big advantage, there will always be a big disadvantage. The right way to think about this is that if you eliminate the disadvantages, you will be left with the big advantages. This is what I call "going beyond common sense," but here again our misconceptions get in the way and it takes a little courage to take a step outside of common sense.

Whether top management, middle management, or the workers who actually do the work, we are all human, so we're like walking misconceptions, believing that the way we do things now is the best way. Or perhaps you do not think it is the best way, but you are working within the common sense that "We can't help it, this is how things are."

Most large companies these days have labor unions, and labor unions are also made of humans so they have misconceptions, and for this reason various things do not always go smoothly when we try to do new things. What is needed is a sort of a revolution of awareness.

Unless we completely change how we think, there is a limit to what we can accomplish by continuing our same thinking. We cannot find a new path unless we take the leap and turn our awareness and our thinking upside down, from top management to the workers, even including the labor unions. Labor unions tend to be good at ideological revolutions, but a revolution of awareness may be a bit more difficult for them.

This revolution of awareness will become exceedingly important. Without it, we are at risk of being happy with achieving only an improvement of 10 percent or 20 percent of productivity as a linear extension of our current ways.

As in the story I told about the *gemba* earlier, it is difficult for people to get rid of their misconceptions that it is cheaper or more efficient to do many parts at once rather than one piece at a time. In particular, when we talk about cost and financial people get involved, they can bring their misconceptions about cost; that it is cheaper to set up and run a batch of 10,000 pieces rather than 1,000 pieces on a press, for example. They think that this misconception is not a misconception and that they are correct, since the math works out.

I received the question "Toyota has been able to make their press changeovers very short. I hear that what used to take one and a half to two hours now takes less than ten minutes, so would it not be more efficient to do the changeover in ten minutes and take the time you made available to produce 20,000 parts instead 10,000 parts?"

You could say this, based on mathematical calculations. If the changeover time took one hour, you would need to run parts for at least two hours. If the changeover time is reduced to ten minutes, and you used the time you saved to make more parts, this should reduce the cost and improve efficiency. The question was whether or not shortening the changeover times and reducing lot sizes reduces the benefit, but this is a completely different way of thinking, so there is really no point in answering the question. All I can really say is, "Yes, according to the math."

CHAPTER 6

The Blind Spot in Mathematical Calculations

When financial people do simple mathematical calculations and think that costs must have been reduced, leaving out the question of the actual quantity that will be sold, this is a large mental misconception.

We produce only what we sell. We often tell people they must not produce what they will not sell, but this seems like nonsense according to mathematical calculations, and people think it costs less to produce 20 than to produce 10.

Perhaps this is very difficult for people to understand. It seems there are many who do not see that just because the results from calculations of strange mathematical formulas are correct, this only means that the answer to that formula is correct and not that costs will actually be reduced.

There are three formulas:

1. Price − Cost = Profit
2. Profit = Price − Cost
3. Price = Cost + Profit

Maybe the financial people cannot understand that each of these formulas means something different.

The first formula is based on the thinking that the product will be sold at a certain sales price. The cost to produce it is subtracted from the price and the balance is profit. You might think the second simply flips the first formula, since it says the profit is the result of the sales price minus cost. The third formula is a bit different, saying that the sales price is the sum of the cost and the profit. When you look at it like this, the formulas may all seem to be the same. Intellectuals seem to have particular trouble seeing that these three formulas have different meanings.

The first formula applies when you are in competition with other firms selling the same product and the sales price is set by a third party, the customer, based on the value of the product. If it takes 80 yen to produce and the sales price is 100 yen, the profit remaining is 20 yen.

The second formula is based on the thinking that we must have a profit of 20 yen, and that as long as we do, things are all right. If the sales price is 100 yen and your cost is 100 yen, this does not leave 20 yen profit. Under this formula, the self-serving answer might be to add gold lining and sell it for 120 yen.

The third formula is not mathematically incorrect. When you move the minus sign to the other side of the equation, it becomes a plus sign, so the sum of the profit and cost is the sales price. However, this means something totally different than the other two formulas. The price is set by the producer at 120 yen because the cost is 100 yen and the producer believes that 20 yen is a fair profit. This fair profit is not gained unless the sales price is 120 yen. Now even if the producer says that this is the correct price, the customer may say, "No fool would pay 120 yen for that" or "Other companies sell the same thing for 100 yen," and so the thinking behind formula number three will not give you a profit if the cost is 100 yen and the sales price is 100 yen.

My interpretation of "cost" in these formulas is that costs exist to be reduced, not to be calculated. The third formula only requires that cost is calculated accurately. In the third formula, profit may be set by what is acceptable to the government so that if the producer needs a profit of 20 yen and the cost is 100 yen, then the price is set at 120 yen, even if the customers are not convinced.

The second formula is the trickiest one. The profit is moved to the other side of the equal sign and the sales price and cost are on the other side. The thinking here is that the costs cannot be reduced, so the value added must be increased, making a profit by producing luxury goods. The idea in the second formula is to shift to producing luxury products. This shift to producing higher-value-added products is a typical philosophy of economists.

The first formula says that the sales price is already set. So the producer must reduce cost, no matter what. The cost that is reduced is the profit that is generated. As a result, if a product that was costing 80 yen to produce can be produced with 50 yen and the sales price is 100 yen, you have earned a profit of 50 yen through your effort. You may be widely criticized if your profit is too great, but I think the first formula is the way to think about cost.

But if you ask a mathematics teacher they will tell you that all three of these formulas are the same, and things will get confusing. Back in 1974 or 1975 an economics professor advised us, "Instead of producing such a high volume of cheap cars and being criticized by the United States, would it not be better to make luxury vehicles that could be more value added and possibly ten times more profitable? You could produce one-tenth the number of cars and still have greater profit."

My thought at the time was, "Economists sure do have a relaxed view of things." His thinking was the second formula, not that the sales price is set by a third party, but that it is more profitable to sell a smaller number of a higher-priced products rather than a large volume of lower-priced products. I suppose you can make an argument for anything and that you can use the same formula to come up with all kinds of ideas.

We at Toyota, and particularly people involved in industrial engineering, use the thinking behind the first formula. We think, "How can we reduce cost?" Costs do not exist to be calculated; costs exist to be reduced. So the most important issue is to try various methods to see which ones reduce cost and which ones do not reduce cost.

In our company we work hard at reducing labor hours. However, many people have the misconception that if you reduce labor hours, you reduce cost. This is a very common mistake for equipment investment, and this is a struggle for us. It is very difficult to persuade people to understand this.

CHAPTER 7

Don't Fear
Opportunity Losses

Whenever we decided to launch a new model automobile, the equipment planners would want to know how many of the new model we will sell. If we will sell 30,000 units per month, then the calculations could be done very quickly. Of course, the calculations are faster if you use computers, but the answer is that if you can sell 30,000 automobiles, then the equipment investment will pay off. That's why the equipment planners said that they could not do production preparation unless we told them how many automobiles we will sell. We don't know how many new model automobiles we will sell.

But the equipment planners said we should know from our demand forecast about how many we will sell. I told them, "If I could predict the future I wouldn't need to come to work. I could make more money betting on horses at the track."

We have been forecasting the weather using highly scientific methods since the Meiji era,[6] but even the weather forecast is not right most of the time. So there is no way we will be able predict people's hearts, tastes, and preferences, no matter how advanced our computers

---

[6] The Meiji era lasted from October 23, 1868, to July 30, 1912.

become. We do not know how many we will sell. If we do not know how many we will sell, we cannot do equipment planning, they say. This sort of foolish talk is troublesome.

If the new model automobile is good, customers may buy 30,000 vehicles per month. But if the new model is bad, and the customers buy them but have bad experiences and other customers do not buy them, we may not sell more than 500 per month no matter how much we advertise. A demand forecast range of 500 to 30,000 is a big problem, the equipment planners tell me. I was telling them this on purpose to make them struggle, and they did think of ways to arrive at a better answer.

For example, we tried prototyping. We see whether the performance of the new product is good or bad. Or we test whether something new will sell or not. We release the new model in small quantities, and if the sales are very good we might expect that it will be a hit with customers. But when this happens the reaction is, "Shucks! We could have sold 30,000 automobiles and made a lot more money but we listened to the old man and we only have enough equipment for 500." So they decide to have enough equipment to make 30,000 vehicles no matter what rather than listen to me. But ironically when you do this, the automobiles do not sell so well and you can find yourself in big trouble.

The Japanese have a tendency of being very afraid of opportunity losses. After the war Japan went through a high growth period between the 1950s and 1960s, up until the oil shock.[7] During this period there was a lack of equipment and manpower so even though there was demand, we could not produce enough. Such a lost opportunity for making money is called "opportunity loss" and is similar in meaning to actually losing money. The word "loss" can be an actual loss or an opportunity loss, and are very different things, but in either case people tend to feel they have suffered a great loss. I think this is another example of a misconception because the lost opportunity to make a profit causes no actual harm, while an actual loss causes financial harm. People confuse these things.

---

[7] The 1973 oil crisis, which set off recessions and high inflation around the world.

Based on our demand forecasts we thought we would sell more automobiles, and looking at our new models we thought they would certainly sell, but the customers did not buy them. This is because our demand forecast was wrong. For automobiles, the sales company creates the demand forecast. You could say, "We can't help it because the sales company forecast is not accurate," but this is not very helpful if your company goes out of business because of this. People are so afraid of opportunity losses that they forget all about actual losses. This is a huge mistake that people fail to recognize, and I guess the lost opportunity to make a profit just sticks in their head.

Just as in the expression "the fish that got away always looks bigger," it is part of human nature that the opportunity that we did not catch looks bigger. The reason that people feel that this is such a big loss is due to another misconception within how we think.

Therefore, it is all right for us to be single-minded about cost reduction when the economy is poor or when growth has stopped or when we have stable growth, to use an expression that sounds better, or when we have low growth. But if we are single-minded about reducing cost when there is no more growth (someone said that the fact that we talk about "negative growth" proves how stuck we are on the idea of growth), when we have zero growth or negative growth (even with a negative, "growth" makes you think you are growing), then this could be the cause of some very bad thinking.

CHAPTER 8

Limited Volume Production Is to Produce at a Low Cost

Lately I hear more news about companies that have succeeded in becoming lean[8] operations or companies that have improved through lean management.

Newspapers and magazines report in the business section from time to time that companies have reduced production volumes or revenues, but have increased profits. However, when you have negative growth that means reduced volume production, and even with zero growth you only have the same volume as the previous year. When the economy is in a period of low growth and the angle on the growth curve is flat, and when you must reduce production and become lean, it is a struggle just to keep costs from going up. When I talk about *genryou*,[9] I use the characters for "limited volume."

---

[8] The Japanese word *genryou* (減量) literally means "reduce weight," and this expression is used both for "dieting" and for companies becoming more streamlined. The English word "lean" was used for *genryou* (減量) management in the English translation of *Taiichi Ohno's Workplace Management*, published by Productivity Press in 1988.

[9] Ohno uses wordplay with *genryou* (限量) by replacing the "gen" character (減) for "reduced" with (限) to mean "limited."

From the standpoint that we only make what will sell and we do not make what we will not sell, it becomes very important for "limited volume" production to be production at a low cost.

You can say funny things like, "It costs less to make 15,000 units," even when you can only sell 10,000 units but what is important is to ask whether the company will make a profit or a loss. Of course, there will be the misconception that it costs less to produce 15,000 units than to produce 10,000 units. There will certainly be situations when it really does cost less to produce 15,000 units. But do you really make more money if you produce 15,000 units and sell 10,000 units while the remaining 5,000 units are moved from here to there and stacked, gathering dust? If you will only sell 10,000 units, produce 10,000 units at the lowest cost possible. It may cost more than producing 15,000 units, but managing limited volume production means thinking of how to produce 10,000 units at the lowest cost possible.

When you have stable growth or low growth it will not be very rapid growth. It may be an increase in volume to 11,000 units. Even then, the important thing with limited volume production is to think of how to produce 11,000 units inexpensively.

The volume (量) of reduced volume (減量) refers to weight, and the volume (量) of limited volume (限量) refers to production quantity. Even with the same word, "volume," there is a difference between weight and quantity. Lean (reduced volume) production does not refer to lowering the production volumes, but rather the volume you must reduce to stay fit. Just as a boxer gains weight when he skips practice and can no longer fight in a weight class, companies can gain weight. In order to fight in their weight class boxers will stop eating to "reduce volume" in an effort to keep their weight down. If a boxer does this poorly he will not have energy and will lose his fights, and this is a very bad way of dieting. The same is true for companies. If companies can get rid of fat when they attempt to become lean, this is good. But if they lose not fat but muscle, and think they are actually slimming down, this is very dangerous thinking.

CHAPTER 9

Reduced Inventory, Increased Work in Process

At a company I visited they told me, "We reduced inventory," thinking that I would praise them if they said this. "We reduced inventory and we have a lean operation," they said.

On close observation, it was clear that they had reduced the raw materials, so I asked whether this caused problems for production and they said, "No, it doesn't." After taking a look inside their factory, I saw that the raw materials had turned into increased work in process.

This was not lean at all. You should not consider raw materials to be inventory. When the materials warehouse is near the manufacturing department, people think that the materials in the warehouse belong to them. When they get bored and they bring some of these materials from the warehouse into the factory and make them into other shapes, this creates a big problem.

Having even quite a lot of material or ingredients does not affect the health of the business. Or, if you know that the price of the materials will increase, you can buy up the material while it is cheap. This is business, so that is obvious. But it is not acceptable for the factory to take this raw material that you should hold onto and machine it however they like and turn it into work in process.

What we mean by limited volume is that we do not make what we do not need. We do not make what will not sell. This is the same for machining or for any process. For example, let's say we have operation #1, operation #2, and operation #3. Operation #1 takes one minute, operation #2 takes two minutes, and operation #3 takes one minute. The processes take different amounts of time. Even if the person at operation #1 finishes one piece, operation #2 takes twice as long, so the piece is still being machined. But if the person at operation #1 finishes the first piece and starts another one, this creates two pieces of work in process. This is how work in process parts can build up.

We cannot sell work in process inventory, and even though we say that we must not make what we will not sell, people think that they have improved efficiency. Or, operation #1 may fall into the misconception that even though operation #2 can only produce 30 parts in an hour, they should work the full hour and make 60 parts. People have the misconception that it is faster, better, and cheaper to make more.

What happens to your company when you make things more efficiently that will not sell in greater quantities? The company will have to pay the people who worked extra hard to produce so much. The company will have to buy twice as much material as they need. Even the electric bill, which would have cost less had the machines been shut down half of the time, will cost more. What happens to your company when you make things that will not sell? Even though the calculation might let you think it costs less, in fact it impoverishes the company.

When we say "things that will not sell" we mean that the downstream process is the customer so that operation #2 is the customer of operation #1 and what the customer does not buy will not sell. Even between operation #2 and operation #3 it is meaningless for operation #2, which takes two minutes and is the bottleneck, to make more parts while the person at operation #3 has gone to the toilet, if the parts they are making are not used by operation #3. This is a very simple thing if people would simply accept it, but when calculations come into play it results in misconceptions. When there are a lot of people who think this way it is no surprise that these companies become impoverished.

Even if the machines are left idle, if there is no demand for production, at least this will reduce wear and tear. Machines that deteriorate in 10 years may last 20 years if they are stopped half of each day. But it would be foolish if instead you used the machines as if you must wear them out and get rid of them in ten years just because they are on a ten-year depreciation schedule. As I mentioned earlier, if someone works very hard all seven hours making holes on a drill press that should only take one hour by making one every 30 seconds, and then he demands a raise because he has worked so hard, the parts he makes will be very expensive. This only increases cost. We need to center our thinking on cost, and instead of just accepting the calculations that show that our methods reduce cost, we need to question whether they really do.

Companies that make several types of parts on one machine will tend to make a large quantity at one time. They start by making a certain amount of part A, then switch over to producing part B efficiently and then to part C. Just as in the example of the press earlier, the calculation tells you that it costs less to produce 10,000 than to produce 1,000 pieces so they keep the machine fully utilized. Then they run out of places to put things. They have no space unless they build a warehouse, so they build a warehouse. Once they have a warehouse they will keep building parts they will not sell just because their calculations tell them they are producing the parts at a low cost. Eventually as both the variety and volume of parts increase, they build racks in the warehouse to hold these parts. And now the moment we've all been waiting for—they install a computer system that will retrieve these parts from the warehouse without error, at the push of a button. Why do they go to such lengths to add cost to the parts they think they have made so inexpensively?

How much does a part that they thought cost them only 100 yen actually cost by the time it is assembled and delivered to the end customer? No amount of calculation will tell you what the true manufacturing cost is in this situation.

Transportation is included in the manufacturing cost if someone in the manufacturing department does it. But it is included in general and administrative cost if a transportation contractor does it, so people think this reduces cost. This type of calculation is nonsense, but people do this openly. These types of cost categories are based on the general common sense of the accountants, and these categories are terrible things. Even though in reality they mostly do not know if the cost was really reduced or if the cost was increased, they say, "That is general and administrative cost, and we can streamline that to make it cheaper. The manufacturing department should just worry about reducing cost." I have heard them say that costs have gone up or down based on their calculations, and I have heard them make logical arguments about how hard it is to streamline the support functions, but this is foolish. It is a big mistake when people are fooled into thinking that costs have actually been reduced by this type of generally accepted thinking.

CHAPTER 10

The Misconception That Mass Production Is Cheaper

There is another widely held bit of common sense that actually comes from a misconception. It is generally thought that mass production is cheaper. Thanks to this, although it is nothing to be thankful about, perhaps there is another bit of common sense that says low volume production must be more expensive. However, when we question whether mass production actually reduces cost, I have to say that I have been around and seen a lot, but there were very few examples where increased production volume actually reduced cost. In most cases increased production volumes increased cost.

What I am saying is that most workplaces have a set production capacity. The machines have a set volume at which the production cost is cheapest. For example, a press may be cheapest when producing 1,000 pieces per hour. Would the cost truly be reduced if this press produced 1,200 pieces per hour?

Of course you cannot buy another press, so the answer is that you work 20 percent overtime. Companies typically must pay 30 percent or 40 percent more for overtime work. When overtime is used to produce 12,000 parts, the real cost has increased by the amount of the labor cost of this worker, which has the overtime premium. However, if a machine that is capable of producing 10,000 pieces only produces

8,000 pieces because there are only orders for 8,000 pieces, then the cost is higher than when the machine produces 10,000 pieces. So although it is true that the cost is reduced when increasing volumes from 8,000 to 10,000 pieces, the cost increases when the volumes must be increased beyond the machine's capacity. Therefore, the cost per piece will be reduced as much as you can increase the production volume, up to the capacity of that *gemba* or factory. However, when you mass produce beyond that capacity the cost increases, by 30 percent in this case.

The unions will protest if we increase overtime, so you buy another machine. After buying the second machine, now the rate of operation is worsened and it costs more, until demand reaches 20,000 pieces. So in reality, there are cases when mass production actually reduces cost and also cases when the more you do mass production, the more cost will increase, up to a point.

During Japan's period of rapid economic growth,[10] customer demand doubled every three years. Those who bought machines looked as though they had the gift of foresight and said, "Our investment in the future was successful." But people are troublesome, in that they remember their successes but not their failures, and so it is not the case that it is always cheaper to do mass production.

And another thing, sometimes you can produce at a lower cost when the production volume is very low. Consider, for instance, a press changeover. When changeover time is reduced from one hour to ten minutes, these low volume products could be made during the changeover time saved. For example, if you had planned for one hour for a changeover and now it takes only ten minutes, you can produce dozens of the low volume part A in those ten minutes. You can then do another ten-minute changeover to part B, produce those parts for ten minutes, and then take the last ten minutes of that hour to set up for the large lot part C. You could make about 50 each of these low volume parts during the time you saved through changeover reduction.

These parts can be produced at no cost, but if we say this, the accountants will scratch their heads. In any case, through hard work

---

[10] From the mid-1950s to the early 1970s.

a one-hour changeover was reduced to ten minutes. So when we ask, "How do we take these time savings and tie them to cost savings?" we should not do the calculation that tells us that it is cheaper to take the extra 50 minutes and produce more of part A. You can take those 50 minutes and produce several parts that each have a monthly demand of about 50 pieces. Again, when you change from part B to part C you can produce low volume parts D, E, and F. This is how you can produce inexpensively at low volumes.

However, as I said in the beginning, because there is the commonly accepted belief that mass production is cheaper, and conversely we accept that low volume products must be expensive, we can sell these products at a higher price. This can be extremely profitable, so as long as the world believes that high mix low volume products are more expensive, you might as well consider this another type of cost reduction and make as much money as you can. If you just use your head, there may be quite a few ways to make money as long as there are all of these misconceptions in the world that have turned into common sense.

To look at it another way, there are still many ways to reduce the overall cost. But if you insist on blindly calculating individual costs and waste time insisting that this is profitable or that is not profitable, you will just increase the cost of your low volume products. For this reason there are many cases in this world where companies will discontinue car models that are actually profitable but are money losers according to their calculations. Likewise, there are cases where companies sell a lot of a model that they think is profitable but in fact are only increasing their losses. This may be very common in other industries, and not just true for the automotive industry.

CHAPTER 11

Wasted Motion
Is Not Work

I have always said that the Japanese language is very well put together, and if the Japanese language is interpreted skillfully it would help the development of Japanese industry. As you know, the characters in Japanese "to move" (動) and "to work" (働) both change in meaning with the addition or removal of the "person" (亻) radical, even though the character is pronounced identically as "doe."

This does not work so well in other languages, since, for example, the English words "to work" and "to move" do not sound at all alike.

I made a similar mistake last year when I went to China. Because the workplace in China was very disorganized with material placed haphazardly, I explained that in Japan we have something called 4S activity that comes from the Japanese words *seiri*, *seiton*, *seisou*, and *seiketsu*,[11] and that we raise awareness of the importance of good workplace organization in a variety of ways, such as by giving awards when the work area is kept neat.

---

[11] *Seiri* (整理), *seiton* (整頓), *seisou* (清掃), and *seiketsu* (清潔) are the original Japanese for sort, straighten, sweep, and sanitize used in the workplace organization discipline known commonly as 5S with self-discipline as the fifth S.

But the Chinese asked, "Why is it 4S?" The Chinese pronounce the characters for *seiri* and *seiton* as "*sei*" just as in Japanese. However, *seisou* and *seiketsu* are written with the character for "pure" (清) pronounced "qing" in Chinese. I was actually aware of this, since we play mahjong and use the Chinese pronunciation for "qing" for the "pure" (清) character in mahjong. But since we were at work and not playing mahjong, we forgot this and wondered, "Why don't they get it?"

We explained to them that using the Roman alphabet, the Japanese words for sort, set in order, sweep, and sanitize all started with "s" so it was 4S, but when they saw the "pure" characters the Chinese could not read it as "*sei*." In the end we said, "Don't worry about it, just remember the 4S." In any case, in Japanese we have two characters, both pronounced "doe," and the one with the "person" radical means "to work" and the one without the "person" radical means "to move."

Because the two words have identical pronunciation, the image of "doe" that is "to move" and "doe" that is "to work" becomes the same. This is a problem. I am not at all joking when I say this, but Toyota City where our factories are located used to be called Koromo City, and in that region they use the Japanese words "to move" and "to work" completely interchangeably. For example, "My wife moves well" means "My wife is hard working," and they actually use "to move" and "to work" to mean the same thing. We built a factory right in the middle of these people, so the employees of Toyota think that moving and working are the same thing. Because they thought that moving with a lot of energy meant they were working, I had a terrible struggle persuading these people otherwise. We should not interpret human motion to mean the same thing as human work. We need to think of motion that includes human wisdom as being something completely different from animal-like motion.

A bear in a zoo will move back and forth in its cage. As far as the bear is concerned this is simply animal-like motion. The children pay the entrance fee to gather around outside the cage to see the bear. In this case, the bear moving its body in front of the children would be an example of working. If the bear is moving and there are no customers watching the bear, this would simply be animal-like motion. So it is very important for people to be able to distinguish between motion that is work and motion that is simply moving, or in other words wasted motion. Likewise, if there is a certain limit to the physical endurance of the bear, it should be kept in the back where it can stay still when there are no visitors to the zoo. By bringing the bear out when the children come to see it, you make them happy so they will want to come back again next Sunday, and this is how the bear can earn a lot of money for the zoo.

The elephant, on the other hand, is brought out in front of the Sunday visitors to the zoo to perform tricks. These tricks are a combination of human creativity and the motion of the elephant, allowing the motion of the elephant to be profitable work. The monkey will be conscious of the visitors to the zoo and move about also, and monkeys themselves actually work, but we do not combine the monkey character to the motion character to write "to work," but we add the "person" radical[12] (亻) to motion (動) to write the character for work (働) .

The Japanese writing system is very convenient in this way. I think the Japanese writing system has contributed a lot to industry by helping people understand both actual and potential misconceptions. At one speaking engagement I was giving this same talk when someone in the audience asked, "Would I write what I do as a combination of the 'water' and 'move' characters?" So I asked him, "What do you do?" and he said, "When I go barhopping I move from one place to another to get a drink. If we write the 'water' radical plus the 'move' character, is it 'barhopping'?" That's his choice.

I often tell supervisors to train their eyes to see the difference between motion without the human element and actual work. Some call this being able to see waste, or asking "How do these motions relate to doing work?"

For example, during the work of machining a part and changing its shape, the worker may skillfully stack up parts five high on the

---

[12] The Japanese writing system, borrowed from Chinese, combines symbols and radicals to create meaning. The person radical (亻) comes from a pictogram meaning "person" (人).

chute between the machines. This is the play of children in kindergarten, not the work of grown men. Companies that pay people for this type of activity will become unprofitable. The supervisor must be a person who can instruct people to not waste motion, and this is the most important role of the supervisor.

We have had the opportunity to see many different companies. When we visit a company, their plant managers or executives will take us around. I don't know if this is standard Japanese, a Toyota City dialect, or a Mikawa area dialect, but they would use the words "the parts were made." This is not the same thing as "we can" or "we were able to make the parts," but saying "the parts were made," rather than "we made the parts." When I asked the managers giving me the tour whether "the parts were made" as a result of everyone in their company busily moving back and forth, or whether they had instructed people to make them, they had no answer for me. If they said "they were made" without knowing about it, they would be admitting they have zero management ability. If they said they instructed people to make these parts, I would have scolded them: "Are you directing people's efforts toward losing the company money?" So they did not answer me. The point is not to have them answer my question, but that a manager's ability to manage will be questioned if they cannot get control of parts that "were made."

The situation where "the parts were made" is surprisingly common. Everyone worked hard and the parts were made. If you asked me, "What is the most important part of production control?" I would say it is to limit overproduction. If you can get away with staring at the floor until the scolding ends whenever "the parts were made," then production control is not doing its job at all.

CHAPTER 12

Agricultural People Like Inventory

The Japanese are descendants of agricultural people and in our hearts there is a sort of nostalgia for agricultural ways. Agricultural people live in a fixed place and grow food in a field or rice paddy nearby. Depending on the weather the harvest can be bountiful or it can be scarce. In countries like Japan where we have typhoons these extremes can be very big, and when there are typhoons or droughts the land is not very productive. So in our hearts there is a way of thinking that we must harvest as much as we can while we can. Even today when modern science has advanced and we work in factories where it does not matter very much whether it rains or whether the weather is good, we still behave as if we were farmers.

When the machines are running well, we think, "Let's all work hard today and make as much as we can, because you never know when machines might break down or when worker absenteeism might get worse." Human beings just cannot seem to get away from the feeling in our hearts that we need to make as much as we can while we can. Although I scold those who build up inventory, even I feel like I want to have extra, and this must be because we have an agricultural mind.

Hunters, on the other hand, can feed themselves when they are hungry by killing a pheasant, for example, and eating it. When agricultural people eat animals, such as when they kill a pig and eat it, they need to store some of the pig meat they did not eat, so they become very skilled at food storage. That is why agricultural people are very good at food preservation and storage techniques as well as at managing this inventory.

So in our gut, we must enjoy inventory management more than production control. Rather than doing proper production control upfront, we prefer to stay busy making things and then later spend effort managing inventory when "the parts were made." Books titled "Production Control" do not sell so well, but when you write books on "Inventory Management" they sell very well. In this respect Toyota has a special reputation for zero inventory or for not having warehouses.

But people tend to understand things in ways that are convenient for them. As a result they build warehouses and then they keep expanding these warehouses. In farming you need to harvest as much rice as you can when you can and keep these food reserves in storage safely away from insects because you never know when there will be a famine. Food reserves that are kept in case of war or famine are meant to be taken out and eaten when needed. However, when you keep stock and hold onto it year after year, this is how inventory builds up.

There is an old farming saying we heard when we were children— "bumper crop poor." When there is a bumper crop, the price of rice goes down. When the fishing catch is very good, in some cases they throw away some of the fish on the shore. When there is a bumper crop or the fishermen's catch is very good, the price can go so low that you might as well give it away. On the other hand, when there is scarcity the price is high, and this is how market price is established.

As market prices were established and as the number of years that these market prices remained in place increased, people started thinking that it would be good to regulate this price to keep it the same, regardless of whether it was a poor harvest or bumper crop. So today we have to pay a high price for rice, regardless of whether the harvest was good or not. In the past when there was such a good harvest that there was more than enough rice, the price would be very low. But now they set a price[13] based on cost calculation, and the price we pay is based on the cost to produce.

Because there is this idea that the sales price is the sum of cost plus profit, these days the rice farmers do thorough cost calculation, including the depreciation cost of their mechanical tillers, the cost of fertilizer per acre, and the yield per acre. Even the daily wage is calculated, although it is odd to talk about farm labor and a daily wage in terms of cost per hour just as we do in factories.

As I said earlier, if you think of cost as something that exists in order to be calculated, the conclusion is that the farmer must receive a fair profit to make it worth his while. This only results in the price of rice going higher and higher. Today there is no longer a market price for rice, but since the Meiji era the influence of this sort of thinking has expanded to all other industries.

---

[13] Japan has a long history of protectionism of their rice sector, and each year the price of rice is fixed by the government.

CHAPTER 13

Improve Productivity Even with Reduced Volumes

These days in Japan, we have quite a surplus of rice. Since having too much rice was a problem and rice production needed to be decreased, a policy of acreage reduction was introduced. Recently rice paddies were also converted to other uses. The government paid farmers to plant things like reeds in the rice paddies so rice will not grow there. As a result farmers have money to spend on trips overseas, and Japan's Agricultural Cooperative tour groups have become good customers for tourist industries in places like London and Paris.

However, even after converting rice paddies there was still too much rice so cultivated acreage was reduced. This meant that the government instructed farmers not to use, for example, 10 percent of the rice paddy cultivated acreage. The farmers were expected to produce 10 percent less rice, but the productivity of the remaining acreage actually increased by 10 percent and the actual production of rice did not change. They said, "We did not reduce enough cultivated acreage," but they must have used pure mathematical calculations and ignored the idea of productivity. They wanted to reduce another 10 percent. They should have instructed the farms to reduce their output by 10 percent, regardless of what acreage they used to produce it, since the point was that there was too much rice being produced. But instead, since they instructed the farmers not to plant rice in some of their rice paddies, the farmers used the leftover rice seedlings to plant rice more densely in their rice paddies. So, of course, the productivity of their acreage increased. This is proof that bureaucrats are not really concerned with productivity. Whether it is the Ministry of Agriculture, Forestry, and Fisheries or the Ministry of International Trade and Industry, they are all Japanese and they all think alike.

The ministries say that structurally depressed industries should scrap the excess machinery. They say the excess machine capacity results in overproduction. But if these companies scrap 10 percent of their bad machines and turn around to replace even half of these machines with new machines that can produce twice as much as the old machines, they are back to doing overproduction.

The top executives in Japanese industry are all descendants of agricultural people, so this makes me think that I should be able to communicate with them. There is no use in complaining quietly about this here, but we need to think more deeply about productivity. It is important to get rid of ideas such as "our productivity now is as good as it will get" and "cost will be reduced if we produce enough volume."

There is a cause and effect relationship between productivity and production volume that just cannot be cut off. There are many cases when the productivity improves because production volumes increase. It may be getting difficult to improve productivity even with reduced volumes. When production volumes increase by 10 percent and you do the work with the same number of people, your productivity improves by 10 percent. Or if demand increases by 20 percent and you increase staffing by 10 percent and you sell 20 percent more, in effect you have increased your productivity by 10 percent. There are people all over the world who are able to do this, not just in Japan.

Is there a way to improve productivity when we have zero growth and production volumes do not increase? In Japan we have full employment and cannot fire people so our hands are tied. It is only natural that when production volume decreases by 10 percent the productivity also decreases by 10 percent.

Anyone can do what is natural. Those companies that can do what is not natural, to improve productivity even with reduced volumes, or the companies that have these people with the eyes to find ways to do this, will be the companies that survive through recessions.

I do not know if this is a good example, but let's say that production volume is reduced by 10 percent. What would happen if we reduced the speed of our machines by 10 percent? The speed of the machines and the rotations of the motors have a relationship with electric power consumption such that the energy consumption is the square of the rotation of the motor. In general, they say that if you reduce the speed by 10 percent you will save energy by 20 percent. Or the reverse: if you increase the speed by 10 percent you will use 20 percent more energy. While this may be common knowledge, we all continue running the machines at the same speed even when volumes have been reduced. When the volume has been reduced by 10 percent, if you can produce enough parts while reducing the machine speed by 10 percent, of course the electric power you need to pay for is reduced as a square of the speed reduced. Through efforts like these you can keep the costs from increasing, even if you keep the same number of people employed.

Although unlike the United States we cannot easily lay people off and we keep our people employed, factories in Japan, depending on their size, have installed many labor-saving machines and equipment. The forklift is a useful device. One of the Toyota group companies builds these products, so we would like people to buy many of them. Even many small companies today have one or two forklifts. Forklifts are advertised as allowing one person to move such and such kilograms or move products so many meters within so many minutes, or that young women can use them to stack items to a great height.

Despite these benefits, if they are battery operated you need to charge these batteries, and if they have gasoline engines they need fuel. If you drive the forklifts the tires will wear. So if you have reduced production volumes you should stop using forklifts and let the people who are idle due to the reduced volumes move materials by hand. Then you will save the maintenance and operating expense of the forklift and this will reduce cost. This cost is much less than labor cost so there is less of a cost reduction than from a layoff, but if you fail to take advantage of these ideas because they are small, you are not being resourceful.

One of these days, even automobiles may not sell so well. We realized this in 1974 so we began thinking about making the transportation lot sizes smaller. The demand for automobiles decreased a little in 1974 and then began growing again very strongly so no harm was done, but in today's environment you never know when production volumes will be cut back. Believing that these years of reduced demand would come again, we began reducing the transportation lot size for materials. You can transport large pallets if you use forklifts, but in a weak economy you should not use them for transportation. Containers should be changed to boxes that one person can transport. The container should be the size that can be transported manually on a push cart. Or, we can use the pallets we have today, but instead of unpacking and placing items on the pallet, we should stack the boxes on the pallet, only as many as one person can transport. We can use the pallets for the moment, and when the economy becomes weak we should stop using pallets and forklifts and instead carry the materials by hand.

Even the pneumatic chucks that close when you turn a valve can be replaced by an idle worker who can turn a wrench by hand. Using pneumatics requires an air compressor, which uses electricity, and if you shut it off, this will keep costs from rising. These types of small efforts will keep costs from going up even with reduced volumes.

CHAPTER 14

Do *Kaizen* When Times Are Good

People involved in manufacturing need to think about making the types of efforts we just discussed. Of course, you could make an argument for shifting production to higher priced, higher profit margin, higher value-added products when volumes are reduced, but I do not think the world works quite that way. If you can do this, then there is nothing better. But there is no sense in just standing by and watching your company become poor just because you do not have these sorts of products, so these are the times that really test your mettle.

If you have not made these types of preparations ahead of time, then when the economy slows down and you find yourself needing to modify your pallets, using smaller boxes and making pushcarts, you will need money to do this and it may be too late. Just as in the expression "the flour costs you more than the rice cakes,"[14] we should prepare in advance about how to operate in poor economic times. This is the real way to do *kaizen*.

---

[14] When Japanese rice cakes are made, a lot of flour is used to keep the sticky rice from sticking to the wooden bowl in which the rice is pounded. Ohno is making an analogy between the cost of the flour to make the rice cakes and the cost of the pallets, carts, etc., to make parts.

Kaizen should be done when times are good or when the company is profitable, since your efforts to streamline and make improvements when the company is poor are limited to reduction in staff. Even if you try to go lean and cut out the fat to improve business performance, when your business is in a very difficult position financially there is no fat to be cut. If you are cutting out muscle, which you need, then you cannot say that your efforts to become lean are succeeding. The most important thing about doing *kaizen* is to do *kaizen* when times are good, the economy is strong, and the company is profitable.

When times are good and by chance our business performance matches up with our cost calculations, we easily become complacent and think, "We don't need to push it." This attitude can result in improvement efforts with no positive results. If you have people move materials by hand instead of with forklifts when times are good, then you will end up adding staff because you do not have enough people, and this is the worst possible approach.

When you have a full employment policy as we do in Japan you cannot have layoffs. Lean efforts based on reducing employment are a problem, and we must protect jobs. For companies that can survive no matter how much red ink they see, it may not matter how many people they employ, but those of us in the private sector must generate a profit.

I will say this again: the only way to generate a profit is to improve business performance and profit through efforts to reduce cost. This is not done by making workers slave away, to use a bad expression from the olden days, or to generate a profit by pursuing low labor costs, but by using truly rational and scientific methods to eliminate waste and reduce cost. I think this is the most important work that industrial engineers can do.

There is an expression "poverty dulls your wits,"[15] which means that when you are impoverished you will not be able to have good ideas and help yourself. When you are impoverished you only think desperate, foolish thoughts. That is why I think we can have good ideas

---

[15] This is similar to "an empty sack cannot stand upright."

when times are good. These days the market conditions are so good for people in some industries that they are laughing all the way to the bank, but I think they may need to stop laughing and prepare for when times are not so good again and step up their *kaizen* efforts.

CHAPTER 15

Just in Time

I have realized this only recently, but apparently the phrase "just in time" is a created expression and not proper English. *Jido*,[16] as in "automation with a human element" and "Just in Time," are two phrases used in the Toyota Production System. I think Kiichiro Toyoda, the first president of Toyota, may have invented this phrase. I have been thinking lately that he took the English words and made them into a Japanese phrase.

According to people in English-speaking countries such as America and England, there is no expression "just in time" in proper English. I heard from one person that "exactly on time" is proper English. Although they say that "just in time" is not proper English, I think "just in time" is a very good expression.

The usage of "just in time" translated into Japanese is "to be just in time."[17] It may be the "in time" that is not proper English. "Timing" is not the same as "time" but rather whether the timing is good or bad, as in whether it is on time or not on time, whether it is "in timing," although I don't know if that is proper English either. The word "just" was added so that enough to be on time would not be plenty in time.

---

[16] *Jido* (自働) means autonomous, as in *jidoka* for "autonomation."
[17] The translation of Japanese ちょうど間に合う into English is, in fact, "to be just in time" or "to have just enough time," so this makes for an awkward translation of this passage back into English.

There is an English phrase "just a moment," which means "please wait a little," and this "little" is not the same "just" as in "enough" or "right on." Being too early is not good and being late is worse, so being "just in time" is good.

For example, if something is needed in the afternoon, it should be delivered by noon at the latest. Delivering it the day before is too early and not "just" at all, but off by a whole day. Anyhow, "just in time" seems like a Japanese-style expression to me.

The correct English expression "exactly on time" implies that you are "on" time, as when you see "on time" displayed for airplanes departing on schedule. We used the "just" from "just a moment" but when we say "just" to native English speakers they would say, "Not just, but exactly or precisely on time."

For example, I understand that foreigners think that Toyota's Just in Time means that if we demand delivery at eleven o'clock then it is delivered right at eleven o'clock. Even Japanese who know English will say, "Toyota's Just in Time is not English" and that "Toyota's Just in Time is too strict on deliveries." There is nothing strict about what we do. Whether the delivery is from a supplier or from another Toyota department, if we need it at one o'clock we require delivery by eleven o'clock. All we are saying is that delivery by nine o'clock is too early and that is not "just."

"Just a moment" means "only a little bit," so you could not say this if you wanted someone to wait three hours or even five hours or you would make them angry. If they deliver parts by around eleven o'clock they will start to run out of parts but there will still be one or two hours' worth of parts available at the line side. It is best if the delivery arrives before we run out of parts. It is best when the next parts are delivered while there are still some parts available. So Just in Time should be interpreted to mean that it is a problem when parts are delivered too early. It is very troubling when some people give Toyota a bad reputation by saying that Just in Time is extremely strict, and that the purchasing department will give you grief if you are even five or ten minutes late. In that sense, I am impressed that "just in time" is a very appropriate phrase whether it is proper English or Japanese-English.

Although I have also heard that during the Meiji era there was something called the Yokohama dialect, which was a mishmash of Japanese and English words, I think this "just in time" is a good Japanese expression made from English words, so if people want to speak proper English and say "on time" that is up to them. If we were to ask an English language expert what the English equivalent of the Japanese expression "to be just in time for" is, the answer might be something quite odd. The words "just in time" are easy for the Japanese to become familiar with and are easy to say. This makes me think that Kiichiro Toyoda was an even greater man than we thought.

CHAPTER 16

Old Man Sakichi Toyoda's
Jidoka[18] Idea

I graduated from Nagoya Technical High School in 1932. This was shortly after the Manchurian Incident[19] and the economy was extremely poor and there were not many places to find employment. As luck would have it Toyoda Boshoku[20] hired me. This company made thread. Cotton thread was the most popular industry in Japan in those days, and as an export-oriented industry the competition was fierce.

Toyoda Boshoku was a very tough employer and for the first three years of employment they paid hourly wages, making me do the same job as the workers in the factory. After the third year I was put in

---

[18] *Jidoka* (自働化) is "automation with a human element," also called "autonomation," or automation incorporating automatic shutoff devices.

[19] The Manchurian Incident (also called the Mukden Incident, or in Chinese the Liutaogou Incident) occurred on September 18, 1931, in southern Manchuria near today's Shenyang when a section of railroad owned by Japan's South Manchuria Railway was blown up by Japanese junior officers. Japan's Imperial military accused Chinese dissidents of this act and annexed Manchuria.

[20] After inventing his automatic loom Sakichi Toyoda established his own spinning factory to produce thread, which he named the Toyoda Automatic Loom Works. In 1918, this company became Toyoda Boshoku. Today, Toyoda Boshoku makes automotive components as well as textile products.

charge of one of the spinning operations, with the title of *kakari*,[21] as in head of maintenance or head of operations, and these were positions just below section manager.

It just so happened that next door to Toyoda Boshoku was a company called Toyoda Automatic Loom Works where they built a machine they had invented. We used this machine to weave cloth. I worked in spinning and had never worked at building looms, but I realized the greatness of the invention of the automatic loom more and more as time went on. Back in those days I thought it was not such a great invention, but I was new and working in entry-level positions so I was ignorant. That great invention was misused by the people of those times. There are some very good things about that invention, but they really did not use it well.

We can say the same thing about the conveyors at the Ford Motor Company, but looking back these highly productive machines were used to make tremendous improvements in productivity using motor power for work that was done with our hands and feet prior to this invention. The most important part of this invention was that there was a device that automatically stopped the machine when the thread broke or ran out. Old Man Sakichi Toyoda called this device "*jido*,"[22] or automation with the human element added.

In the past, motorized looms were powered by belts turned by pulleys attached to one large motor or engine. As small motors such as the five-horsepower or three-horsepower motor were invented, these motors were attached to the loom, one per machine. We used to call these "independent motors." I think these somehow developed into the automatic looms, without the human element. Each loom has a power source and would keep spinning thread as long as the motor was running.

The critical part of Old Man Sakichi's invention was that the machine stopped when the thread broke, because when the thread

---

[21] *Kakari* (係) means "in charge of" or "head of" something.

[22] Automatic is pronounced *jido* (自動), and the word *jido* (自働) of the same pronunciation was created by Sakichi Toyoda to mean "self-working" or autonomous. The "human element" refers both to the イ symbol, meaning "person," that was added to change the meaning, and to the autonomous ability to stop.

breaks, the loom produces defects. Even if the thread broke, a machine will keep on spinning defective cloth if no action was taken. This cloth would be defective product. Although we now understand "automation with a human element" to be the avoidance of making defects, back in those days people only saw that it improved productivity tremendously and used it as a way to whip workers into making more product. They saw only that they could not make money unless they hurried to the machines that had stopped to connect the thread again.

Since the machine would stop by itself and avoid making bad product, they should have thought about how to make thread that would not break. Instead we saw young women running to a machine that had stopped to quickly connect the thread so that the machines could quickly start again and they could continue producing.

It sounds bad to say that Old Man Sakichi's invention of the automatic loom "with a human element" was abused for making a profit because business performance improved by making people work harder, but those were my thoughts at the time.

Even among the spinning machines they implemented *jidoka* with a human element here and there, but these were also used to make more parts and to improve business performance through the efforts of the workers.

During the war Toyoda Boshoku was merged with Toyota Motors and became the textile department, and after the war the textile department was returned to a company by the old name, Toyoda Boshoku. I did not work with automobiles during the war, but on airplane parts. I was working in a drawing process for making brass tubes used to cool oil in heat exchangers for airplanes. After a year or so of doing that, I was moved to what is today the Honsha plant of Toyota Motors. It was about six months before the end of the war, in February, when I went to the Honsha plant of Toyota Motors. I was assigned a position equivalent to section manager for the final assembly plant.

Looking at things in this way, there was a very strong mind-set of relying on labor to build automobiles. Back in those days we thought that we needed to think of ways to increase production while using the same labor by implementing a bit more *jidoka*.

CHAPTER 17

The Goal Was Ten-fold Higher Productivity

Around 1937, when I was still working in spinning, a plant manager from Mitsubishi Electric, I believe, returned from visiting factories in Germany and America. I sat next to this person by chance on one occasion, and he told me that between Germany and Japan there was a three to one difference in productivity. When he asked at the German factory how many people they employed, expecting 1,000 or so, he learned there were 300 people. When he asked about the production output of this factory, it was three times what he expected. These were the reasons why he said that the difference between Germany and Japan was three to one.

After he saw this in Germany he went to the United States to go see another company in their industry. Then he told me that the difference between the United States and Germany was three to one. This is how I heard that there was a productivity difference of nine to one between the United States and Japan. Back in those days the machinery in the United States probably didn't use transfer machines, and even in Japan we were not using domestic machine tools but machine tools from the United States or Europe, so there could not be so much difference in the equipment we used. Nevertheless, the difference was nine to one and this was something I could not imagine.

It may have been that after the war had ended and some years had passed, some amazing machines were used in the United States to increase productivity many times over, but immediately after the end of the war I think the machines and equipment in use in Japan and the United States were not so different. After the war when the Allies landed, someone from GHQ[23] whose name I forget announced that the productivity in the United States was eight times greater than that of Japan. At the time I thought that since I had heard in 1937 or 1938 that the difference was nine to one, the Japanese must have made some improvement during the war now that it was eight to one, but in any case this was an almost unimaginable gap in productivity.

The automotive industry is a flagship industry for the United States, and since the "eight times greater" number given by the GHQ was probably an average, we expected that we would need to raise our productivity ten-fold higher in order for the Japanese automotive industry to be competitive, so we thought of various ways to improve our productivity ten-fold. We realized that with our traditional methods the Japanese automotive industry would never survive. We could not even catch up unless we improved our productivity ten-fold, and this is why we have been working on how to improve our productivity ten-fold for many years.

However, at the time there were no companies working on such drastic productivity improvements. If we went to see Ford or GM, we might understand how, but it was not as though we could implement what we saw right away. We could not succeed unless we totally changed our thinking. That was the spark that led to the Ohno System.

In those days we did have a conveyor in the final assembly plant and the assembly work was done in a flow. An old-timer in the assembly section told me how President Kiichiro Toyoda said to him that the most efficient way to assemble parts in an assembly plant was when each part arrived Just in Time.

Until then, the upstream processes would bring the parts they finished one after another. For example, when the engines were

---

[23] GHQ stands for General Headquarters, the base of U.S. occupation in Japan after World War II.

finished they would push them into the assembly plant. However, because we lacked steering wheels, for example, or because we could not gather a set of parts, we had what we called the "intermediate warehouse" at the time. The way things were, the intermediate warehouse was full of materials, but we could not build automobiles out of them easily. The parts were good finished parts, but a car is not a car unless there is a steering wheel. Or we may have everything else but the engine. Because of these sorts of things the assembly area could not easily assemble cars.

The reason was that the upstream manufacturing processes would make whatever they could, and all parts were being produced in an out-of-control condition, so we could not run the assembly line until the 17th or 18th of the month. By the 17th or 18th most of the late parts would have arrived. Just as in the expression "the sumo wrestler does his work for the year in ten days, and he's a lucky man,"[24] we often said in those days that we did a month's work in ten days. So even if we look at only this, we could not have finished the month's work unless we had three times as many people as we needed since we were doing the work of 30 days in 10 days. That is why we thought we could do the work with one-third of the workforce if we could effectively use so-called *heijunka*[25] to level out the assembly workload between the 1st and the 30th of the month.

Because I started out in assembly, I realized that we should be able to do the work in assembly with one-third of the people as long as we could gather the parts properly. So as far as improving productivity three-fold, it looked like there was nothing to it, since we could triple the productivity using Just in Time to gather the parts. We tried this and learned that we were in a condition to easily improve our productivity many times, depending on how we looked at things.

---

[24] In the old days the sumo wrestler worked only ten days per year. Today the sumo wrestler has 60 days of wrestling matches per year, in six tournaments of 10 days each, one match per day.

[25] *Heijunka* is the averaging of both the product mix and the product volume in order to create a smoothed or level production schedule. It is often called production leveling or production smoothing in English.

By making various small labor-saving improvements, it is not so difficult to improve productivity further to about five times the original.

However, for labor-intensive work like assembly that must be done by hand, it is hard to improve productivity no matter how much *kaizen* you do. But the people who are doing the work feel that they are being made to work harder since now they are assembling the same number of vehicles per month with one-third of the people. What I mean is that although previously the assembly line started and stopped during the first 20 days of the month and did not run properly at all, the people thought they were doing work. The Japanese are very diligent so they think, "We have no parts. The line isn't running. I might as well do some cleaning over there." They have the misconception that they are working just because they are using their labor, even though they are not getting work done.

Everyone mixes up motion and work. That is why when people look at just the results and see that the output has tripled, they think that the productivity was improved by making people work a lot harder.

CHAPTER 18

The Supermarket System

So that is how I came to work with automobiles, but the work we did in textiles with *jidoka* and the efforts to find ways to improve overall productivity turned out to be very useful for me. What I mean is that in those days at Toyoda Loom Works one person would operate 20 or 30 machines. This meant that as long as the machines were running, things were good. So people had to run to the machine that had stopped, reconnect the thread, and start the machine again as soon as possible.

When another machine stopped, they would go running again to connect the thread and start the machine, so one person would run 20 or 30 machines, depending on the type of textile. In effect, when the machine was running the person had nothing to do. And when the person was doing something, the machine would be stopped, so we thought that if we changed our mindset about automotive work, one person should be able to run ten or more machines. If one person ran ten machines the productivity would increase ten-fold. Strictly speaking this is not actually so, but this type of thinking was at the very foundation of the Toyota System.

Back in 1951 or 1952, the first of our classmates to go to the United States came back with all sorts of color photographs he took, the type that you display with a slide projector. Among them were several photographs of a supermarket. He explained that in the United

States there was something called a supermarket, and there was only a young woman at the exit, and the customers pulled along something like a baby stroller, bought just what they wanted, and paid at the exit. This reduced expenses quite a lot, so the customers could buy things inexpensively. If one person was enough for the store, this reduces expenses for the store. Hearing this, I thought that they should be able to make a profit even if they sold goods at a low price, and this may have been one hint for the idea for having the downstream process retrieve the parts. That is why when we first started this around 1952 or 1953 we called it the supermarket system.

In those days the Japanese were suckers for English words, so rather than calling it the Ohno System or the Ohno Line, we called it the supermarket system for quite a while. Eventually we saw that this idea of having the customer go to the store to buy, or the downstream process—the person who wants it buys what they want in the quantity they want—was identical to Just in Time. We realized that this was also the system with the best productivity for the buyers as well, since the buyers can buy according to the size of their refrigerators and the amount of money in their wallets, and live economically.

If you think about it, the traditional Japanese grocery sales methods such as home delivery service and grocery men walking their neighborhood routes may seem very convenient for customers, but in fact these methods increase cost.

For example, even with tofu, in the morning when the tofu was ready the tofu seller would walk around playing a flute, selling tofu. Not only was the tofu very fresh, it was brought right to your door. But on the other hand if you waited, thinking to add tofu to your miso soup tomorrow but on that day the tofu sold very well so there was no tofu left by the time he came to your door, you would need to go out in a big hurry to the store. Although it appears convenient at first glance, the Japanese were actually living an uneconomical lifestyle.

On the other hand the supermarket is the total opposite, since the customers drive their cars to go shopping. This is the total opposite of Japan's service-oriented spirit. If you order home delivery and you only need two stalks of long onion, you can hardly ask them to bring two stalks of long onion, so you order a whole bunch. You think, "I might as well buy some daikon[26] too," so in the end this is an uneconomical way of shopping. Whether in Japan or in factories the principle is the same. As the parts are made, they are taken to the user. Although this may seem like service, when the parts are brought just because they were made, whether you need them or not, assembly areas are forced to work in a very uneconomical way.

If you do this well, increasing your productivity about three-fold is easy. That is why starting from the idea of Just in Time we end up with a system in which the upstream process only needs to produce as many parts as the downstream process takes away. For every ten parts that are taken away, simply make ten parts by the time they come back the next time.

---

[26] Giant white radish.

CHAPTER 19

Toyota Made the
Kanban System Possible

At first the supermarket system was something like a method for the upstream process to replenish the materials. Whatever the assembly area took away, the upstream process had to produce before assembly came to get those parts again. The *kanban* was a slip that indicated how many pieces they were coming to get, so that if they were going to take ten parts this became a production instruction slip directing the production line to make ten pieces. The machining operation has various conditions such as lead times, lot sizes, and so forth. So the big question for the upstream process becomes how to effectively produce only what the process downstream came to get.

In order to do this, they must work to perform machine changeovers in less time, even if it is only a little bit less. It is a problem if it takes one hour to produce something but things do not go according to plan, and if something else is needed, then it is no longer Just in Time. Therefore, you need to make lot sizes small. When lot sizes are small, you need to do changeovers more frequently. We realized that if changeover times were so short as to be negligible we could deliver Just in Time. This is what required us to pursue single-minute changeovers.

However, in those days the workers would not go along with such outrageous ideas. What I was particularly worried about was the support of upper management for such a risky, unproven approach that was off the beaten track. Upper management would normally have been too afraid to give permission, but I think one of the big forces behind the development of the Toyota System is the fact that Chairman Eiji Toyoda and the late Advisor Shoichi Saitoh let me try this to my heart's content.

If I had not been at Toyota Motor Company, I think another company would never have let me try this. So Toyota made the completion of this system possible. Today it is called the Toyota System, but it was around 1961 or 1962 that this name was adopted. Before that, because it was so risky and we were afraid that one mistake could lead to the company going out of business, we called it the Ohno System.

What I mean is that back in those days when we tried to get the people on the *gemba* to try something, they would ask, "How does Nissan do it?" or "What other company is doing it this way?" so I would tell them that there may be nobody else doing it this way, or that someone may be doing it, but that I had not seen it so we were going to follow the Ohno System. I worked on this as the Ohno System until the mid-1950s, aware that one misstep and I would have to pay a very steep price for failure. During that time there was certainly quite a lot of resistance from various quarters, and the top managers did not understand. The workers themselves understood even less. Everyone was worried whether or not what we were doing would put the company out of business. However, we kept going with the belief that the Japanese automotive industry would not survive unless we saw this through to the end.

In those days the main goal was improving the productivity per person, and this is basically true even for the Toyota System today. It was only after the 1973 oil shock, when we no longer knew if we could produce in large volumes, that we began using the term "lean"[27] production.

---

[27] The Japanese term *genryou* (減量) literally means "reduced weight," "dieting," "slim down," or "reduced volume." When used in the context of management, the meaning is similar to lean management.

Prior to that, for many years we had been operating on mass production alone so we could improve productivity just by increasing volume. All we had to do was to increase output ten-fold with the same people, so that was very easy now that I think about it.

Just in Time led to *kanban*, but in those days we were not thinking of *kanban* at all. The basic principle in our traditional method was for the producer to deliver to the downstream process. There are quite a few companies that still operate this way today. We just reversed this process so that the downstream goes to get what they want from the upstream process, and this was Just in Time. So there was no struggle here, none at all.

Just in Time is an ideal, and the common sense in those days was that there was no way we could do such a thing. That's why I coined the phrase "beyond common sense" later on, and maybe this is because I tend to go against the grain and it was part of my personality to look at things backwards.

When I was a child I was often scolded for being contrary because I would always look at things from not just from one viewpoint but from several viewpoints, questioning things from various angles. According to the month-end rush production viewpoint, the upstream process makes parts and puts them into the warehouse, and they think they are doing their job. They say, "We got our work done," but meanwhile the final assembly line has not been able to build even a single automobile. They may have hundreds of engines, but if they have no steering wheels it is no good. If you are missing the rear axle, you cannot build an automobile. If you do not have the frame, you cannot build an automobile. Perhaps starting out in the assembly area was a good education for me.

When we actually started working with the *kanban* system in the mid-1950s, I was in charge of only a part of manufacturing, so I only had authority to make changes in my area. In the 1950s I was in charge of three areas, which were the machining plant, the assembly plant, and the body plant. What we called factories in those days we would call departments today. Another person was in charge of the so-called raw material–forming departments such as casting, forging, and heat treating.

Therefore, I could only use the *kanban* system between the machining and assembly factories or the stamping and assembly factories. When I became the plant manager of what is today the Motomachi plant in 1962 or 1963, I was placed in charge of forging, casting, and the so-called material-forming departments, and we were able to use the *kanban* system at the whole Motomachi plant for the first time. Until then, because different people were in charge, it was difficult to go to other areas and get them to use the *kanban* system. Before I became the plant manager of the Honsha plant, I was at the Motomachi plant from 1959 or 1960 until 1963. While at Motomachi we did things like *kanban* wherever we could, but because raw materials came from the Honsha plant and for various other reasons, we could not use *kanban* in the Honsha plant or in all areas of the Motomachi plant.

The *kanban* system is something that affects all departments. Of course our workers were hesitant, but we also had to be careful to minimize confusion for our suppliers who made and delivered our parts.

This is an extreme example, but when I was at Motomachi we used *kanban* in select areas only and the purchased parts were the very last to be put on a *kanban* system. If you cannot get your *kanban* system to work properly within your factory, there is absolutely no way you can use *kanban* with your suppliers.

When I returned as the plant manager of the Honsha plant, another person became the plant manager of the Motomachi plant. At this time, they implemented a *kanban* system with suppliers in one area and caused extreme problems for the parts suppliers. At that time I scolded them, and told them not to take half measures with *kanban* and forbade them from using *kanban* at Motomachi for a while.

CHAPTER 20

We Learned Forging Changeover at Toyota do Brasil

When you actually implement Just in Time production you need to reduce changeover times and as a result make the lot sizes smaller. The most difficult area to do this in is forging. The casting area is not so difficult. People just have the misconception that it is more efficient to make the same parts continuously.

In hot forging processes the metal is heated until it is red hot, placed in a die, and hit by the dies. Because heat is used, the die must be heated to a certain temperature. The raw material needs to be cut to the right size and heated, but not overheated so that it will melt, nor can it be heated too little so that the machine is idled. There is also an oxidization layer or scale that forms during the forging process. This scale flies about and makes the adjustment of the dies on forging presses very difficult.

In general, during a die adjustment, the top and bottom dies are attached and one part is hit. The dimensions are measured and the locations or the height of the dies is adjusted. The part takes the proper shape after two or three pieces are hit. This is why the forging changeover time takes the longest.

It also takes a long time from when a die is changed until the parts are finished. That is why calculations show that unless you run forgings in large lots you will lose money. This is why forging was the last area to be addressed through changeover reduction.

At around that time in Brazil,[28] we had purchased one piece of forging equipment in order to bring the manufacturing of forgings in-house. We had only one machine and we needed to make more than 60 different types of parts. In Brazil we only built two or three cars per day, or about 40 per month. They must have been the smallest automobile company in the world, although these days they build around 400 per month. Because of the low volume, nobody in Brazil was willing to supply Toyota with forgings.

It is common knowledge that forging suppliers will not run your parts unless the order size is more than 1,000 pieces. Who knows when the supplier would fill orders of two or three pieces? So with more than 60 varieties of parts, if we ordered 1,000 pieces we would be buying years of inventory. If our daily demand was 1 or 2, or 40 per month, and we ordered 1,000 pieces, which they might not even run for six months, Toyota in Brazil would lose money for sure. We realized this would not work, so that is why we installed one forging press and told them to produce all of their own forgings for all 60 types.

However, we told them they must not make more than ten pieces at one time. We told them that if the changeover takes one hour and they did eight changeovers in eight hours, they would have no time to make product. That is why you must think of ways to do changeovers in 15 minutes, we said. If you do a changeover in 15 minutes and produce parts for 15 minutes, you can produce two types of parts in an hour. That means you can make 16 models in a day. We found that they could easily run through all 60 models in one week, and this is how I had Toyota people study forging changeover at Toyota do Brasil.

The supervisors at Toyota do Brasil were Japanese, and this worked out well. When we said something with conviction, it made the people

---

[28] The forging changeover activity in Brazil was in the early 1970s. The metal press stamping shops had 15-minute changeover capability at Toyota by 1962. Forging changeover reduction took a decade longer due to the nature of the process.

believe it could be done. If we had said the same thing at a Japanese company the reaction would be, "Sure it sounds good but that won't actually work." There is a huge difference between being told to do something in Japan but believing it cannot be done, and being told the same thing in Brazil by a bossy Japanese man with a mustache and believing that it can be done. That's how they made it possible to do changeovers in less than ten minutes. However, we did make some suggestions such as doing what is called "external setup," having the dies prepared ahead of time so that they could be rapidly exchanged and parts could be produced only minutes later. Because the scale burns onto the guides, it was not possible to place guides on a forging press. Without these guides we simply could not make the first piece a good piece. So we had to figure out how to attach the guides.

We made it possible to retract the guides when the forging press was hitting parts. We thought if we set the heights properly during external setup, attach the die to the press, and retract the guide, then we should always be able to make the first part a good part. Although we made the dies in Japan because we could not afford to hit two test parts when we were only making ten parts, the guide pins were all installed in Brazil.

Toyota do Brasil did a very good job and they were able to produce more than 60 varieties of parts without causing even one part shortage.

As a result, the roles were reversed and we sent two or three people from Japan to Brazil to learn about forging changeover. After that they became a lot more active with this in Japan.

On the other hand, casting is not so difficult. Castings are made by pouring molten metal into a mold. So it should be normal to produce different parts one after another, yet people have the misconception that it is more efficient to use the same mold to make the same part over and over. There are also differences in material properties. The castings may be made from malleable or just regular pig iron, so if the lot size of the molten metal is large, you have to cast a certain quantity of the same part. Because the material is different, melting smaller lots requires extra effort, time, and heat, and this is undesirable.

However, when we are talking about a company the size of Toyota Motor Company, even smaller lots would be 1,000 or 2,000 pieces. In that sense, the higher production volumes have made Toyota less concerned about changeovers.

As I mentioned earlier, at Toyota do Brasil they produced a low volume but a wide variety of parts. At Toyota do Brasil they thought of a solution for melting materials in common in a larger lot and then using additives to make just the quantity of the particular material they needed.

From that perspective, Toyota do Brasil was a model case or a test plant for high-mix low-volume production and the place where the Toyota System was implemented the best. At Toyota Motors' production volumes today, the good thing is that you do not even need to do changeovers.

There are lines that are dedicated to making certain parts for certain models, so to be honest there is very little need for changeovers these days. There is less and less need to do a press changeover after five or ten pieces. This makes me think that Toyota do Brasil might be doing the Toyota System better than any other location. However, although for their low volume products they are not able to reduce the cost as much as a mass production operation could, Toyota do Brasil is making a lot of money producing volumes that you would normally expect to cause them to lose a lot of money.

Back when we started with the Toyota Production System, we would have demand for 3,000 to 5,000 vehicles per month, and a lot of variety. It was not so-called high-mix low-volume, perhaps it was medium volume, though there were some low volume items. So the Toyota System is a system that works very well when applied to midsized companies. What I mean by this is that the Toyota Production System was born in the days of 2,000 to 3,000 vehicles per month, so when production volumes are as high as they are today at Toyota, you do not really need to use this system to reduce cost.

CHAPTER 21

"Rationalization"[29] Is to Do What Is Rational

When I was implementing the Ohno System, it was difficult to get people to understand what I was saying. In these situations it is true that from time to time I would say, "If you can't do what I say, get out of my sight." But I also said, "Just do it. Don't worry. I'll take responsibility."

The way the factories were in those days, if I gave instruction to the department head and waited for him to tell the section head and then for the section head to tell the team leader, there would be no knowing when things would get done. Given this situation, although I did not give instruction directly to the workers I did skip over all of the layers and give instructions to the frontline supervisor. Partly, I was impatient. Also, by the time my instruction was passed on verbally across 10 or 12 people the result would come out quite different.

Likewise, even when I gave instructions through the managers they would make their own judgment or solution, and by the time it reached the front lines it would be something very different. It was

---

[29] In the context of business, rationalization (合理化) means "streamlining," or becoming more fit through *kaizen* activity.

not easy to get my will across. Sometimes I would listen directly to what the foremen had to say. That was how I abused my authority.

That is why there was quite a bit of protest from the middle management, but the individuals who were personally asked by the plant manager would be motivated to try things. However, it would be bad if this person's direct supervisor had not heard anything about what they were trying, so this person would have to tell their supervisor that I had asked them to try something. There was no need for them to tell me what they were doing or what the results were. This is how we educated our managers, both from the top down and from the bottom up. So although I would give instruction directly to people, sometimes I would scold them if they reported the results directly back to me and not to their direct supervisor. Even this direct supervisor should not just relay my instructions. If they have ideas of their own they should aggressively try them. I did let them know that if I went to see what they were doing and did not like what I saw, I would get mad at them.

Nobody knew if the Ohno System would work. Nobody else was trying it. If the results were good, that was good, and if the results were bad we needed to change right away. I would go in the morning and tell them to try this and that, go at noon to see that they had done it as I had instructed, but if the results were poor we had to change right away. Sometimes it would indeed be "the morning's orders are revised in the afternoon." I had to say this over and over to the engineers, because engineers are so hardheaded. They stuck to their ideas, so I reminded them that "the wise mend their ways" so they should change and become wise men.

If other companies had been trying the Ohno System I would have taken our people to see and understand it, but probably nobody else was doing things like we were. Even if we could see and copy what another company was doing, if we did not change it further we would only be as good as the company we had seen.

Back in the days when we would visit Nissan or when Nissan would send people to see what we were doing, I would get reports saying, "This is what Nissan is doing," but I would tell them that we have to think of even better ways. Because Nissan was doing better than we were at the time, our people would first try to copy them. This was not good, so we stopped sending people to see Nissan. At some point Nissan stopped letting Toyota see what they were doing.

Before the war Nissan had purchased an American factory and moved it all over to Japan, and there were even engineers from overseas who came along. So in that sense, Nissan must have been much more advanced than Toyota. During and toward the end of the war, the Americans went home, and soldiers would come with green bamboo[30] and prod us, "Make more, make more." After we lost the war the military inspectors left, but those habits stayed with us. Although we knew a lot from a technical standpoint, the Japanese stopped taking orders easily from other Japanese.

In 1956 or 1957, I saw my first American factory, but saw that what they were doing was ordinary. There was nothing fantastic about what they were doing. I was able to see factories of GM, Ford, and American Motors but it was all very common sense, or should I say that when a production line has been rationalized there was nothing extraordinary about it. The more rationalization efforts progress, the more it appears they were only doing things that are obvious, from the point of view of the third party. When something looks fantastic there must be something bad about it. So if you tour a factory and think "Wow!" then this is not such a good factory. When you see a factory and think, "There is nothing worth seeing here," they may in fact be doing a lot better.

It is not easy for the Japanese to do what is reasonable without putting up any resistance. The simpler it is, the harder it is to do. Rationalization means doing what is rational so there should be nothing that makes you think "Wow!" For example, the easiest way to

---

[30] By "green bamboo," Ohno is very likely making a reference (possibly humorous) to the fact that during World War II Japanese citizens were urged by the military to take up arms by making spears out of green bamboo to prepare for the invasion by Allied troops. Children in school were taught the proper use of the bamboo spear. It is very possible that Japanese army soldiers came to the factories with green bamboo spears during the war to demand production for the war effort.

move round things is to roll them, or you can make a heavy thing lighter by putting rollers under it—these things are obvious.

When something has been completely rationalized it should be in a simple condition, but everyone makes rationalization too complicated and this is no good. It is odd to say "reduce inventory and work in process" in the name of rationalization. If you rationalize, there should be no work in process. If you only need one and you have two, this is not rational.

CHAPTER 22

Shut the Machines Off!

The symbol for "work" (働) in the automatic loom[31] (自働織機) has the human element (イ) and is not the same as "move" (動). No matter how much bad product you make, this is not considered working. So the idea of *jidoka* is that when there is a defect it is a good thing for the machine to be shut off. This means that you do not make defects. If you make defects, you have not worked. That is the difference between autonomation with a human element and automation without a human element.

The idea of stopping comes from the notion of autonomous, or "self-working," machines. There is a definition of this, which is "*Autonomous machines are machines which have automatic shutoff devices.*" I had this definition in my head because I heard this definition by chance during my student days. I was in mechanical studies, but I had several hours of classes in spinning and weaving. During one of these classes we were taught the definition of "autonomation." Therefore "work" means having automatic shutoff devices.[32] In today's terminology, these are sensors and such, but I

---

[31] Although it is called the automatic loom, the Japanese is written as 自働織機 or "autonomous loom."

[32] Ohno is saying that the automatic shutoff device is what differentiates automation (自動化) and autonomation (自働化), or self-moving (動く) and self-working (働く).

believed that these things had to shut the machines off. This idea remained in my head until it was useful to me decades later.

The teacher for the spinning and weaving class at the Nagoya Technical High School had studied the autonomous loom because Toyoda Automatic Loom Works Ltd. was located in Kariya, a city near Nagoya. So they were probably not teaching this definition at other schools. When there were no automatic shutoff devices, the workers themselves would have to shut the machines off. It is harder to use sensors effectively in assembly factories. In these situations the workers themselves will push the shutoff button and a regular assembly conveyor belt can become an autonomous conveyor belt.

One time when I was visiting another company to give them instruction, their president complained that all I taught them was how to shut things off and asked me to teach them how to make things. However, this cannot be avoided so we teach the workers to stop the line when they find a problem because when they stop it, we can take countermeasures. So the first step toward autonomation with a human element is thinking of a way to stop the machines. How can we detect the defects and automatically stop the machines? That is why the final assembly line at Toyota is a conveyor that has the human element built in. Although the line does not automatically detect the defects, the workers themselves will stop the line if they are conscious of quality.

That is why there are a lot of shutoff buttons on the assembly line. The person who finds the problem stops the line. Stopping the line creates a great loss, so this forces us to think, "How do we keep them from stopping the line?" and this results in more and more quality *kaizen*.

The thing that foreigners who visit Toyota plants find most puzzling is the fact that workers on the assembly line are stopping the line right and left without hesitation. This is not done anywhere else in the world, and even in Japan it is very rare outside of Toyota. The first thing we did at Hino and Daihatsu, which are Toyota group companies, was to install shutoff buttons. Then at Daihatsu instead of so-called "lines that we stop," we are making them "lines that stop themselves" without a person pushing the button.

So, we need to think of how to keep people from stopping the machines. When we were first trying this out on the assembly line at the Motomachi plant we told the workers, "When you become tired, stop." Then we would ask the team leader or foreman why that worker was being made to do work that made them tired. When workers found a bad part attached, they would stop the line and we would think of how to prevent the attachment of bad parts. And last, I told them over and over they had to think of ways to make it so that even if the workers wanted to stop the line, they could not.

Things like safety and quality are fundamentals. Everyone notices that making defects just increases cost. So reducing cost must be at the very basis of quality control. That is why around 1955 or so when we needed to reduce cost, the first thing we did was to work on reducing defects. Cost is reduced when you reduce defects.

CHAPTER 23

How to Produce at a Lower Cost

When Toyota began building passenger cars, the Crown cost more than 1,000,000 yen.[33] Mr. Ishida, Toyota's president at the time, was summoned to the Japanese parliament and received criticism that Japanese automobiles were too expensive.

In those days some influential people thought we should not be building automobiles in Japan at all, but that we should buy them inexpensively from the United States, so cost reduction was the most important task for us. Japanese automobiles were expensive, and we had to make them cheaper.

In 1949 or 1950, right about the time when Toyota was going bankrupt, a man named Ichimada, who was the governor of the Bank of Japan, was saying that we should not make automobiles in Japan, and that it would be better to buy them from the United States.

Because of this situation it was an absolute directive to reduce cost as much as we could. We knew from before the war that building passenger cars in Japan simply did not pencil out. So first off we thought if we could increase productivity—for example, a ten-fold

---

[33] The exchange rate was fixed at 360 Japanese yen to one U.S. dollar between 1949 and 1971.

improvement in labor productivity—we could match the United States in terms of labor cost.

On the other hand it was not possible for us to lower cost through mass production because in those days we did not have the customer demand. Toyota nearly went bankrupt in 1949–1950. We nearly went bankrupt because we built 1,000 cars per month but could not sell them. That gives you an idea of how weak the Japanese economy was in 1949 and 1950. When Toyota put together their reorganization plan, they thought they would not be able to sell 1,000 vehicles per month. The reorganization plan was based on less than 1,000 vehicles per month, 940 if I remember correctly. Right about when the labor disputes ended and we began rebuilding, the Korean War started. We were saved because Japan received special orders for vehicles from the U.S. military.

Before that even if we built 1,000 vehicles per month there were not enough customers for them. Whether this was good or bad is a separate issue. So 3,000 vehicles per month was like a dream. The volume was less than 1,000 and, of course, we could not build 1,000 of the same model. We built perhaps 60 passenger cars per month according to our reorganization plan. It was no wonder that they scolded us for selling them at 1,000,000 yen, but when I think about it now I can hardly imagine it.

We could not just build thousands of passenger cars because if we could not sell them we would go bankrupt. But in order to export the Crown to the United States we needed to build about 1,000 vehicles per month. We were told by the president of Toyota Motor Sales, Mr. Kamiya, that we needed to build about 1,000 vehicles per month. So how could we make it possible to build 1,000 per month? This was back in 1955 or so. These days we build 1,000 vehicles in a matter of minutes, so there is no comparison.

The Toyota System was very effective during these start-up years, but later our production volumes grew and grew, overshadowing the impact of the Toyota System. So the Toyota System remained unknown to the outside world. It was only after the oil shock in 1973 and 1974 when production volumes fell and Toyota was profitable in spite of this that the Toyota System began to attract attention.

Without those tough times, the Toyota System may have become more of an American system. What I mean by this is that since one vehicle model will sell tens of thousands of units, you can do model changes every three years, and you can continuously invest in equipment and still not have enough capacity.

After we had converted to the *kanban* system, when people from Daihatsu came for practical training to Toyota, I told them to only make what they needed. One of the trainees said, "We still have materials. I still have time so isn't it better to make as many as I can?"

I replied, "No. That is not right. We only make what we need. If we need 100 but we make 120 pieces just because we have materials, this is a negative for the company. If you need 100, make 100. But you should run many processes so that it takes you all day to make 100."

Once during the war we got into a lot of trouble when everyone went home at noon because they had finished their scheduled work for the day.

One key point in the process of how to produce only the quantity needed at a lower cost is to give each person enough work. If people finish their work at three o'clock and they still have material they will think that it is a waste not to keep making more. However, the lowest cost method is to make each person responsible for more processes so that if the demand is 100 pieces it takes them all the way to five o'clock to finish 100 pieces.

One of the main fundamentals of the Toyota System is to make "what you need, in the amount you need, by the time you need it," but to tell the truth there is another part to this and that is "at a lower cost." But that part is not written down. That may be why people misunderstand and think we go home when we've produced the quantity needed for the day. The Toyota System is to make what we need in the amount we need, at a lower cost.

I have been telling people that they must not put "at a lower cost" first. There are all kinds of ways to produce at a lower cost. If you can make 120 in regular hours but you only need 100 so you only make 100, this increases cost.

The most difficult thing about the Toyota System is to study how to produce these 100 pieces at a lower cost, so everyone must study how Just in Time enables you to produce at a lower cost.

If you put "at a lower cost" first you can make various mistakes such as overproducing or not making enough, or getting the timing wrong. There is no end to the pursuit of the Toyota System and how to produce at a lower cost.

During this process, it is a challenge to balance the work effectively. One area may be idle after three o'clock while another area may have two hours of overtime. You must absolutely avoid ending the day early because your scheduled work is done, because this will increase cost tremendously. But this is what will happen if you do not balance the work.

The key point you must study about the Toyota System is how to produce at the lower cost during limited volume production. If the "limited volume production" aspect is ignored, and instead you think that mass production will reduce cost, and you think it is better to make 200 instead of 100 just because you have five people available, this goes against the Toyota System.

CHAPTER 24

Fight the Robot Fad

"Limited volume production" is an expression we began using in 1973 after the first oil shock. Until then, we could sell as many as we could build, so it was relatively easy to reduce cost through mass production. However, in 1973, automobile production temporarily dropped a little. There are many industries whose volumes have kept dropping ever since. But no matter how much I tell this to people in the automotive industry, they just don't understand. They are not thinking about slimming down at all. To be blunt, I believe to this day that automotive companies work as if costs going up or down do not matter as long as they can get the orders and produce in large volumes.

Due to international trade friction, exports will be restricted starting this year.[34] When this happens, we cannot just say that we will make up the difference in domestic sales, because it is not as if domestic demand will grow indefinitely. This is why unless the automotive industry takes the question of limited volumes and how to produce at a lower cost seriously there will be companies who live to

---

[34] Japanese automobile manufacturers agreed in May 1981 to a voluntary export restraint (VER) program in response to criticism that the Japanese market was closed to American automobile manufacturers. This program limited exports to 1.68 million Japanese cars to the United States each year. This cap was raised to 1.85 million in 1984, to 2.30 million in 1985, and abolished in 1994.

regret it. As far as lowering cost through producing large volumes, they all have experience with that.

The very foundation is "how to produce at a lower cost" and this is essential. It is wrong to skip over this foundation to pursue high performance using robots or automation. I suppose people install robots in part because it is a fad. They install robots to keep up appearances, or to say they have reduced man-hours. I wonder whether many times they do this without thinking too much about whether the cost was increased.

It may seem like we are opposed to robots, but whether we are talking about robots or computer systems, progress is necessary. But we must be careful not to ignore the question of how much cost was reduced when we implement computer systems or robots. Perhaps I am being a nuisance to robot manufacturers by saying this. There is an English person who said that robots and automation ought to be internationally banned by the year 2000. If you consider this, it may be good to get them while you can, in case you cannot when you need them.

When I said in China that they should not modernize, they objected, asking why should China not use robots, when in Japan there were many robots in use. They have so many people in China, why do they need robots? Using robots solely in the name of modernization is not right.

Even in Japan today, robots and NC[35] equipment are selling very well, but the important question is whether these things are really reducing cost. In the days when you can sell everything that you produce, it is okay. If it was not limited volume production, there are probably many situations where using robots would certainly reduce cost.

Certainly man-hours will be reduced when you use robots. If you calculate manufacturing cost based on man-hours, the production cost will be very low. Although in situations where robots are used to do work that is very dangerous it is unavoidable for costs to increase,

---

[35] Machine tools that use machine-readable sets of numeric codes to control their operation are called NC for "numerical control."

I think the mind-set of wanting to install robots because they are a fad is a bit odd.

We really should use more automation and robots, but the simplistic use of automation or use of robots is a problem. This is a very important point. It is wrong to automate something just because we can. I have been saying this at Toyota Motors for many years, and these days at Toyoda Gosei I am giving them an earful. There is a sequence for implementing automation that must be followed, even though it is hard. Automation just for its own sake is a problem. Automation should come as a result of a need, but if the need is mistaken, it can be for appearances' sake. Just as pianos were a popular fad at one time so that people wanted a piano in their house because the neighbors had one, or people thought, "I'm a section manager, and I have a daughter, so it looks bad if I have no piano," and were proud of their piano, regardless of whether there was musical talent in the family, buying equipment to keep up appearances is wrong. This does not matter as far as the people selling pianos are concerned. They just want to sell pianos.

Cost reduction is the number one need for using robots. Second, robots may be used to further the respect for people in situations where the work is dangerous, even if the cost is increased to some extent. However, even for dangerous or unpleasant work there are cases where robots cannot do the work and humans must do the work. But it is very bad to ignore the human aspect and make people do dangerous work just because it would cost more to use robots. The problem we need to watch out for the most is the use of automation as playthings for engineers and *kaizen* experts.

When you really do not have enough people, cost analysis will show that it is better to use robots than to use people. But with our aging society, what will happen to cost when robots do the work and our aging workforce become the "watchers" of the robots?

There are people who say that in Japan as the wages increase, as the society ages, and as there are fewer people in the labor force, there will come a day when we will need to use robots. People have different ideas and viewpoints on this topic. For example, those in the Ministry of Labor are concerned that robots will take the jobs of workers and displace people. The very fact that they are thinking about this seems odd to me, but it goes to show that whether it is the use of robots because of a fad or whether it is robots taking away the jobs of people, these are issues we need to think about. In the United States, for instance, the workers are very vocal, and they demand wage increases. Robots just work and do not say anything, so even if the unemployment rate is high, more robots are installed. This may be good for the company but what does this mean for society as a whole?

CHAPTER 25

Work Is a Competition of Wits with Subordinates

In order to lead a large number of people, you have to be tough when it comes to work. However, I think this is basically not a matter of giving orders or instructions, but a competition of wits with subordinates. I tell people, "When you give an order or an instruction to a subordinate, you have to think as if you were given the order or instruction yourself." And if you lose this competition of wits, you have to swiftly admit it.

However, most superiors will state their wishes in the form of instructions or orders. They do not think for themselves at all, but instead say, "You are the expert so you should be able to figure this out," or "This is how I want you to do it." If your reaction is, "Oh, I see," when the subordinate comes back and says that it is just not possible, you should not be giving orders. It is no good to say, "I have broad responsibilities. I am busy and I have no time to get involved in this. It's your area of expertise so you think about it."

You need to struggle together and think about the problem together. You have to offer various suggestions, as much as possible. If the subordinate comes back and says they tried what you said but it did not work, you must have some advice to give them, or you will lose their respect. When this happens and you do give them a suggestion, and this time it works, you must say, "My suggestion was poor but you did a good job figuring this out." The important thing here is to present an attitude that the subordinate will understand.

As a form of self-improvement or mental training, I think we should work on becoming people that others will follow. But this is a tricky area. You have to have a different attitude with some people. Everyone has a different personality. Even if you say the same thing, some people will not listen or do what you say at all, while some will.

"Follow me," is not an easy thing to say. But when people do follow you, take care of them through thick and thin. Once again, this is mainly about the *gemba*, but when we were implementing the Toyota System, I would say, "Get out of my sight if you won't do what I say." But I was supportive and grateful to those who followed me. People will not follow you otherwise. People know there will be good times and bad times.

CHAPTER 26

There Are No Supervisors at the Administrative *Gemba*

I have spoken about the "*gemba*," but you can view office work the same as the production floor where we make things. You can have the "*gemba* philosophy" for administrative work by identifying your administrative *gemba*. At the administrative *gemba*, nobody is watching. While it is easy to see what is going on in production areas, it is very hard to see what is going on in managerial departments. When you see a person diligently working on something, it is difficult to judge whether this is really something that must be done right now. The trouble is that at the administrative *gemba* the managers do not have the mind-set or the wits of "supervisors." They need to have supervisors. Supervisors must, of course, be able to teach.

There is also the question of how you evaluate the results of work. The supervisor in Japan traditionally did not supervise the work. Too many of them would supervise how the people were working. Whether the factory or the office, they all make this mistake. The supervisor must supervise the progress of work. However, what we find is that supervisors are supervising how the workers are moving. This is one area that definitely requires *kaizen*.

From our point of view, being a supervisor is more difficult. Managers can get by on knowledge alone. Supervisors need knowledge, of course, but more than that they need to be able to demonstrate. In other words, they need the ability to teach in order to be a supervisor. Managerial departments think they do not need supervision. They cannot see the work.

I am making a lot of noise these days at Toyoda Boshoku, telling them not to be deceived by the appearance of work. They say, "Look how hard our people work," but I tell them, "That is not called working. They just have fast hands." If you rely on workers with fast hands, you will fall behind in automation. Instead of trying something different, they think they will be more efficient if they make the young women work faster. They are busy supervising the motions, for instance, noticing that one young woman has slow hands, and do not have eyes to supervise the work.

Even in professional baseball, the supervisor, or the field manager, has a huge influence. Someone who has never played baseball would not make a good field manager, but likewise the person who was a very good player in their youth is not always a good field manager. The field manager needs to know each of his players well. Unless the field manager has a firm grasp on various aspects of his players, including personality, skill, and so forth, he cannot truly lead the team with authority. When the field manager has a winning game plan but a player does not perform, getting angry at the player for not being good enough still will not win them the game. In a similar way, it is important to develop strong supervisors in the production and office areas.

However, white-collar workers who are on an "elite" career track tend to get rotated out of their position. That is why nobody is looking at what is happening with the work. After a few years of doing one thing, they move on to another position. Nobody really checks to see what they did while they were there.

This is true of both supervisors and managers, but there is no effort to see how much better they can do than their predecessor. This is a bureaucratic mentality that expects to be promoted to a section manager after serving a certain number of years without incident, and then after a few years as section manager to a department manager. Nobody looks at them and asks whether this section manager did the same work as their predecessor with two or three fewer people.

Toyota Motors is no good at this either. At Toyota, efforts to rationalize the office work are liable to end up making things less efficient. The true impact of rationalization should be measured by the number of people who work for the section manager or department manager and the total amount of work they have accomplished, but they do not measure things this way.

Do managers have the mind-set to look at the work their predecessor did with 50 people and say to themselves, "When I am head of that section, I will get the same work done with 40 people"? If a manager is in charge of the same section for a second year, they still do not measure their work by saying, "Last year we did it with 50 people so this year we will do it with 45." If it is just, "Things went well last year and things went well this year," that section manager is not making any progress. When the answer to "So why does their salary keep going up?" is "Thanks to the union," this is a problem.

CHAPTER 27

We Can Still Do
a Lot More *Kaizen*

We have done a lot of rationalization on the *gemba*, and we are near the limit. In contrast, it's becoming common sense that administrative processes still need a lot of rationalization, but we can still do a lot of *kaizen* on the *gemba*. However, the impact of *kaizen* will get smaller compared to the effort required.

That last bit of improvement may or may not give you 1 percent improvement, despite all of your effort. That is why *heijunka* becomes very effective, but relatively few people realize this. There is a lot more we can do with *heijunka*, and by considering the production sequence we can improve things.

Let's focus on assembly. Normally the manufacturing departments should think about making the quantity needed in assembly. Assembly requires a variety of parts. The individual processes making these parts think they have increased their productivity. They may look at their cost calculations and think that there is no more room for improvement. But if they only need 100 pieces and they are making 120 pieces, and they are saying that rationalization helped them improve productivity, this is not *kaizen* at all. They should be thinking about how to make 100 pieces with even fewer people.

The important thing is to produce in sets. The calculations will show that producing in sets costs more. The actual product that you ship somehow costs less, but the individual parts cost more.

They say, "We have all of the engines we need for this month. The parts were made. Our efficiency has greatly improved. The production floor is running well. But we have no differential gears." So they cannot build an automobile. They should build one set of differential gears while they build one engine, and a front axle, a steering wheel, and one frame. The smooth coordination of this type of sequence must be the job of production control.

Looking at things individually they say they are doing a good job of producing gears, or that they are using robots very well, or that they can do the work with just three people. But these items can only be sold when they are together as a set. Producing "the things we can sell, in the amount we can sell" is a very simple idea, but there is nothing so difficult as actually doing it.

When you think about producing in sets you can see what a bad thing overproduction is. The problem is that the calculations do not show this.

Making too much or making the right amount too early is beyond waste. When you take the point of view that this harms the company, we can still do a lot more *kaizen*. If you rely on calculations, things will be very rational, but you cannot expect much of a positive impact from rationalization.

It seems hard for people to understand that the company will profit more if you make 100 pieces than if you sweat to make 120 pieces. Of course, if you can make 20 percent more than you need, you should be ashamed that you are carrying too many people. When you look at administrative work from this perspective, it is quite bad.

CHAPTER 28

Wits Don't Work Until You Feel the Squeeze

When I'm sitting in a place like the chairman's office I have no idea what is going on at the *gemba*. But when I used to go to the open office of the production control department to see what people were doing, there was a young woman there who would always make a telephone call whenever our eyes met. If the person on the other end of the telephone was free, maybe that was okay, but if they were working, the call is a nuisance to them. When our eyes met, she felt awkward and had to make a telephone call to keep up the appearances of being busy. This was a terrible thing that she was doing. But she thought she was doing some work. She thought it was okay since I was watching from a distance so I could not tell who she was calling or what she was talking about. She may have thought that she was just passing time, but this also wasted the time of someone outside of that office.

When I say "*gemba*" it is not just the production workplace but also the office or any workplace. When you observe the *gemba* there are many things you should notice, but unfortunately there are many managers who just think they have to keep people from being bored. If they can prevent people from getting bored and complaining, they have succeeded. They are not questioning at all whether the work that is being done now is really needed. They do not know how to question things and use their wits.

Earlier I spoke about a "game of wits," but just think of your wits as something that does not work unless you feel the squeeze. So how do we make everybody feel the squeeze?

Humans, and this is true of other animals, will absolutely use their wits when they are in trouble. We use our wits in various ways that we have found to make our daily lives more convenient.

How do we make them feel the squeeze? In order to make them feel the squeeze, you have to feel the squeeze yourself, so that you can use your wits also.

Specifically, the game of wits is to think of how to make people feel the squeeze. If you can make them feel like they are being squeezed to death, they will come up with good ideas for sure. But managers don't make people feel like they are being squeezed to death. The reason for this is that if the person who gives them a problem does not have any of their own ideas, the only reply they can give when their subordinate comes back and tells them, "It is impossible," is to say, "Oh, I see." Then the subordinate is off the hook. If you want your subordinate to feel so squeezed that they believe saying "It is impossible" is not an option, you must feel the squeeze and struggle just as hard with it yourself when you give your subordinate the problem.

I suppose this means that we have to become attractive people.

Most of the people working in the automotive business were men, so the question was how to attract other men to me. When you are attracted to someone or when you have fallen in love with someone, even their faults can appear charming, so you have to think of how a man can attract other men. There is no formula for how to draw other people to you, so you just have to make a sincere effort…. On the other hand, men do seem to study how to be attractive to women. We need to improve ourselves and make efforts to become people whom others will follow to the ends of the earth.

The young women at Toyoda Boshoku are very charming and kind. At my age it is not as though I am doing anything to attract these young women to me, as the attraction between men and women is a different kind of attraction. But it's true that whether it's men or women, when people are drawn to you they will volunteer to get things done for you. The important thing is to show your face frequently. That way, when I am on the *gemba* they will talk to me more readily. Or if the woman is bored when our eyes meet, it would be much better if she smiles at me instead of making a telephone call. People will not change if you just tell them this. If you don't want them to make the telephone call when your eyes meet, you need to think seriously about how you can get them to not make the telephone call.

CHAPTER 29

Become a Reliable Boss

I never get angry at the workers. However, with supervisors and above I will get very angry. The *gemba* is a convenient place to get angry at people. There is a lot of noise so they can't really hear what I am saying. When I scold the supervisors on the *gemba*, the workers see that their boss is being yelled at and they sympathize with their boss. Then it becomes easier for that supervisor to correct the workers. If you call the supervisor away to a dark corner somewhere to scold them, the message does not get through. The *gemba* is a noisy place anyway, so if I am yelling at them and the person being scolded doesn't really know why they are being scolded, this is okay. However, when the workers see their boss being scolded and they think it is because they are not doing something right, then the next time the supervisor corrects them, they will listen.

When I first became a supervisor and manager, one of the old-timers told me of the boss named Jirocho of Shimizu[36] who would never scold his subordinates in front of other subordinates. He would take them off to a corner and scold them. I was told that you must not

---

[36] Jirocho of Shimizu was a legendary big boss of Japanese outlaws in the mid-1800s. He was famous for protecting the weak and standing up to the strong. He became a folk hero for his deep compassion and the loyalty he displayed and inspired. He was said to have over 1,000 followers and subordinates.

scold supervisors in front of the workers, but I was doing the opposite on purpose. If you really hit a nerve when you scold them, they will get angry. But if they are being yelled at in a loud voice and they don't even understand why they are being yelled at, I think this takes the pressure off of the person being scolded. Even though everyone hates being scolded in front of their subordinates at first, in the end it makes it easier for supervisors to communicate with their workers.

Even at the *gemba*, things will not work out this way unless you have worked together for a long time. In order for the *gemba* to be a place where the people directly doing the work can work with vim and vigor, they need to have someone they can rely on. From that perspective, it is better not to change team leaders so often. When someone is put in charge of a department and their area of responsibility is increased, people will depend on them. But if after a year or two this person is assigned to another factory, the workers lose their reliable boss. I feel sorry for blue-collar workers. They need reliable bosses. If the reliable people are frequently moved to other positions, they can no longer be relied upon.

On the other hand white-collar workers each think they are on their own path. In that sense they have no reliance on each other. That's why when they reach retirement age we have to find them another place to work.[37]

When the bosses change frequently, people think, "Even if I work hard for him, I will be left behind when he becomes a plant manager someplace else." They don't know who will be the new boss. When this is the mood, the *gemba* loses its vim and vigor. It is a bad thing when you create a mood among your workers of "The boss will be promoted if he just gets by each day. We're just workers anyway." There is a huge difference in productivity over the long haul between people who work with vim and vigor and people who are just putting in another day at the job.

---

[37] It is common practice in Japan for senior managers who reach retirement age and are obligated to retire but wish to continue working to leave the company and join another company, typically a supplier, in an advisory or consulting position.

One time I gave an engineer on staff who worked on the *gemba* a hard time. I summoned him, and a young woman from the office went to tell him the plant manager was calling for him. So the engineer came running right away. I said to him, "If I really needed to see you I would go to the *gemba*. If you can come running to see me that means the people on the *gemba* don't rely on you. If people in the *gemba* did look to you for help, and if they were coming to you with their problems and you were thinking hard about these things, there would be no way that you could come running just because the plant manager called for you." I told him that if it was truly urgent, I would go see him myself. "You should not be standing at attention here in front of me, having worked up a sweat running to see me. You're a disgrace. The people on the *gemba* don't rely on you at all." I said to him.

If you are out there observing at the *gemba*, do something for them. If you do, the workers will think, "He's watching us but he comes up with some good ideas." That way when the workers see you they will look forward to your help again, and as a result they will begin telling you what makes the work hard to do and ask you to think of ways to make it better.

If the workers think, "There he is again just standing there. He must have a lot of time on his hands. He never does anything for us," then nobody will come to you with their problems.

I said, "When you enter the factory you should walk in a way that takes you hours to go 100 meters. If it takes you no time at all to walk 100 meters that means no one is relying on you."

Instead of just correcting workers that are sitting down and working when they should be standing, you should find ways to make the work easier for them. This might be letting them sit if they have many tasks, or teaching them how to use a tool properly. When you do this, the word will spread quickly. People will say things like, "That person came and did this and now my work is much easier." Once that happens, people from other areas will also come to you for help, and it will take you a while to walk 100 meters when you go to the *gemba*.

CHAPTER 30

Sort, Set in Order, Sweep, Sanitize

"Sort is to throw out what you don't need and set in order is arranging items so that they are ready when you want them. Arranging things neatly is lining things up, and proper management of the *gemba* requires sorting and setting in order."

At the *gemba*, the words "sort and set in order" are very commonly used but most factories have not actually achieved this.

I used the expression *seiretsu*[38] when I visited a particular *gemba*, and not only was there no first-in, first-out system in their warehouse, but their warehouse was accepting whatever their suppliers brought to them. There were parts that were no longer usable because of design changes, but since "the parts were made" they never got rid of them. It took a very long time for that factory to bring the parts from their suppliers to the assembly area. I told them to sort and set in order, and when I visited them the next time, the parts were lined up in neat rows. I asked them, "Don't you know what it means to sort and set in order?"

---

[38] *Seiretsu* means "forming a line" or "standing in rows."

Sort means to throw out what you do not need, as when you do personnel adjustment. If you are holding on to your parts and stacking them up in your warehouse just because you worked hard to make them, this is not sorting.

For setting in order, it is just as it says in the characters for *seiton*. In the old days there was a medicine called *tonpuku* (頓服), which worked immediately after swallowing it. The character for *ton* is the same. Setting in order (整頓) is to arrange (整) things so that they can be immediately (頓) retrieved. If you need to move everything out of the way to retrieve the one item you need, you have neither sorted nor straightened. Things are just lined up in neat rows.

People who have been soldiers would say that arranging in neat rows is lining up. When soldiers are told "line up" they will do so neatly, in two rows, for instance. But if the orders were to gather and set in order, they would not know what to do. Marking lines to prevent people from going in and out from an area or to limit the height for stacking things has nothing to do with sorting and setting in order. I told them, "We call sort, set in order, sweep, and sanitize the 4S, but we do not have an S for 'lining things up.'"

At Toyota Motors they also used to just line things up. We had a "4S Competition" where once or twice per year the executives would go to each workplace and give awards for areas where sorting, setting in order, sweeping, and sanitizing were done well. In the early days it was a "Lining Things Up Competition." But this was not good since the first item received was buried at the bottom and they would have to move everything on top of it out of the way when they needed to use it. This was not setting in order at all. After that Toyota Motors got better and better at true sorting and straightening.

If you are not careful, sweep and sanitize can just use up a lot of paint. The goal is not to make it clean in the sense that it is colorful and pretty, but clean as in "sanitary." That is what is really needed. Even with sweeping, of course, you need to clean away the chips and dust in a machine shop, for example. However, after the war, partly due to the influence of the United States, "color coordination" was a fad for a time and we did things like painting machines the same color. But that type of "clean" is different from the clean we mean when we say "sanitize."

Another reason for this confusion is that it sounded better to make it 5S. The words sort, set in order, sweep, and sanitize all start with the letter S, but discipline is very important.

The word "clean" can mean several things, but in the sense that the work environment is improved and that people can feel good working there, sweep and sanitize will not work unless people are motivated. If people are not concerned because they know there is a cleaning crew, the factory will always be dirty no matter how much you clean.

Self-discipline was added at the end, and in general, discipline is like education, but there is something more than that. So in reality it is difficult to do, but if you do not keep discipline in mind, once things start to get sloppy they will get worse and worse. For example, golf courses are very particular about etiquette, but no matter how loudly they remind you, people don't do as they say. However, if they give up because people don't listen, things will get worse and worse. Particularly, there is less and less discipline and too much emphasis on academics in schools these days. Because of the nature of athletics, athletes, on the other hand, are disciplined by other veteran athletes or their teammates.

It is a problem when athletics are left out and the only emphasis is on academics, because there is nobody there to teach discipline. These days there is a lot of violence in the junior high schools, and I think the reason for this is that the kids grew up without anybody teaching them discipline. When a large number of people are working together and in groups, things get harder and harder unless we each keep our discipline.

Discipline should be taught by those near to you. If you try to get everyone together to teach discipline, this is what we used to call "morals training" in elementary school. We no longer have morals training in schools, so the parents should teach discipline, but instead both parents work and we have latchkey kids. Since the schools do not teach morals, there is nobody to discipline our children. Not only that, these days there are parents who were never taught morals in school. Their children will be taught no discipline at all. Rather than the strict morals training of the old days, various people, from parents to classmates, should learn morals and ethics each day in a variety of ways. You will not learn true discipline if you get everybody together to read a book for one hour, but then do not follow up.

We need somebody to nag us about these things. And the person that does the nagging definitely must practice what they preach or else they will lose credibility and people will say, "What the heck are you doing?"

These days there is a trend for people not to pick on discipline, and that is the biggest problem. Discipline is taught when seniors scold juniors. This is not only in work but also between elders and youth. There must be scolding and correction, in both directions, and not just talk but followed by action.

CHAPTER 31

There Is a Correct Sequence to *Kaizen*

"Manual work *kaizen* means thinking of better ways of using the existing equipment. Rather than making tools (equipment), it is important to think of how the work should be done."

The idea is to think of even better methods of using the existing equipment, but equipment is also advancing rapidly these days. There are times when new machines are installed, and when this happens, you must think of even better methods for using those machines.

If you install a robot and think only, "This is convenient. It does my work for me," then this is not an example of using the robot effectively. It is important to do *kaizen* on the robot as soon as you install it, or to change the method to work well with the robot.

Instead if you say, "We can do this if we could just buy a robot," and buy a robot without even knowing if there is a better method than the current one, then eventually you will not do *kaizen* at your *gemba* unless you can buy a robot. That would be bad. But if you first see how well you can use the machines you have, and how you can use the existing equipment even better, then when you install the latest machines you can add on your improvements.

The more you install the latest pieces of equipment, the harder they are to use.

Back in my student days at the beginning of the Showa era[39] there was a camera called the Pearlette. It used a single-element lens that that did not let much light in and the shutter speed was also very slow. This was a camera you could buy for 10 or 15 yen. In those days there was a model called the Leica that cost 300 yen. But the Pearlette took good pictures. The models like Contax that had a lot of peripheral equipment were actually harder to use.

The machines you have today are Pearlettes, so if you say the machines are old and you give up, *kaizen* will end. On the other hand, a person who cannot even take decent pictures with a Pearlette may not take good pictures with a Contax camera and its peripherals, such as a telescopic lens and wide-angle lens. In fact, it may be harder for this person to take good pictures with the better equipment.

Likewise, the more that the equipment is the latest machine or a high-performance machine, the harder it becomes to use. But if you master the use of the equipment and can take good pictures with a Pearlette, you will certainly take better pictures with a Contax. The type of people who say, "I can't use the Pearlette. I don't even want to take pictures unless I use a Contax," will not be able to take good photos even with the Contax.

These days you have the instant camera or cameras where all you have to do is push a button, and these are very inexpensive so it's hard to go wrong. In reality, when you want to take the right picture, you need the right conditions.

Rather than saying that you could take a good picture with a Contax but not with the camera where you just push a button, we should question whether the person who cannot even use the simple camera would be able to take a good picture with the Contax. This is hard to say.

---

[39] The Showa era lasted from 1925 to 1989.

And another thing about *kaizen* is that there are various types of *kaizen*, three or four if we are to talk about "manual work *kaizen*" and "equipment *kaizen*" and "process *kaizen*." I have not explained myself enough, but the *kaizen* we are talking about here is "manual work *kaizen*."

I will say that first you must do manual work *kaizen*, and then equipment *kaizen* and process *kaizen* following that. So as you can see, there is a correct sequence to *kaizen*.

Do *kaizen* to the manual work first. Once you know how to make more improvements, but the machine you have is preventing this, then you can look at the right machine that will further improve productivity and quality.

When the latest machine is installed before you have the skill to do *kaizen*, the machine will end up running you. There is a correct sequence to *kaizen*.

People who do not have the skill to do *kaizen* will say, "This machine will do the work of five people with only one or two. Our efficiency will improve if we buy it. The calculations show that the investment will pay off," and so they want to buy more and more machines. This is a problem.

There is a related story from back when Toyota Motors' Kamigo plant was just built, and also at the time when the Motomachi plant was newly built. Workers from the Honsha plant of Toyota Motors were sent in separate groups to the Kamigo and Motomachi factories. The people from the Honsha plant who had done *kaizen* effectively on their old machines, and had trained in good methods, now had only the latest machines to work with at the Kamigo plant. The plant was built later, so this is to be expected. These people modified and did *kaizen* to the new machines again right away. Even if the new machines were built a certain way, the people found a better way, and they would change it.

But if you have a new factory and you hire new workers and use inexperienced workers to run the latest machines just because they can be run by inexperienced workers, the machines end up running the people. As a result, it becomes impossible to see which way actually reduces cost. You need the ability to continuously modify and improve your current equipment. You may know how to operate a machine, or what things the machine tool manufacturer can teach you, but those are just basics.

The type of people who say, "Unless they buy us that machine there is no point in doing *kaizen*," will not be able to do *kaizen* no matter what kind of machine you give them.

Next is process *kaizen*, and an example of this would be to make a big improvement by reversing the process sequence. In manufacturing you have steps, such as do this at process #1 and then do this at process #2.... What can be very clear is when you have the inspection process at the very end. Process *kaizen* would be to inspect the product during the process so that every piece was a good piece. Another example is to find the defect in the process and prevent it from being passed on to the next process. In an even simpler example, most factories do inspection after the product is complete, meaning that their final process is an inspection process.

But if you believe that quality is something that is built in at the process, what you can do is to inspect at each process, and in some cases you may not need your final process at all, or you may need to only check a few things at the last process.

Likewise, the more you look at the steps in a machining process and question whether moving this step in the cutting sequence before that one will improve efficiency, the more you will see that there are many areas to do *kaizen*.

It is only natural that each person inspects their own work. When there is a cutting step that must be performed first in order to avoid causing shrinkage porosity at a later step, as long as this is the first cut, the product will be a good product. But if you do not detect the porosity until the very last process and then you reject the part, this is a huge loss.

In that sense, if process *kaizen* is left up to the engineers, and the machine operators just absentmindedly follow the process the engineers set up, then you also lose the value of manual work *kaizen*. In some cases you will find as you are doing manual work *kaizen* that certain steps can be done at a different time, combined, or done at the same time.

The first step in *kaizen* is manual work *kaizen*. Manual work *kaizen* is the most important, because as a result of manual work *kaizen* you will learn many things about changes you need to make to your equipment, or changes you need to make to your process as a result of changes you make to your equipment.

When people think of themselves as process *kaizen* experts or as being in charge of equipment and do *kaizen* efforts separately, even the motivation for manual work *kaizen* will be lost.

Related to manual work *kaizen* there is something called multi-machine handling, which is when one worker operates multiple machines. This is a fundamental principle of the Toyota System that requires you to clearly separate machine work and human work. Doing human work means doing work that can only be done by humans.

Taken to extremes, this means doing the same work all day long. The thinking behind this is that when the machine is doing the work, a person does not need to be there. This is where the idea of multimachine handling started from. Take the example of a lathe machining on automatic cycle. Just because a person is standing and watching does not mean that the machine will be more honest, or that when the person walks away the machine will act up and begin machining improperly. The person loads the part, turns on the switch, presses the automatic cycle button, and the machine will do the work. No matter how many minutes the person watches the machine running, this is not work. If the person has time they should be loading and unloading the next machines. This is how humans do human work and machines do machine work. The first thing is to make this distinction clear. In general, people still mix these things up.

People will say, "This takes five minutes," so I ask, "How much of the five minutes is human work?" They reply, "It takes 30 seconds to unload the part and 30 seconds to attach the next piece and push the button. The machine is cutting chips for the other four minutes." They would say "five minutes" because in the past there would be one person per one machine. The "man-hours" were five minutes. They always thought of the machining time and the man-hours as the same thing. But in fact, the machining time was four minutes. The manual work time was one minute. But they say, "This takes five minutes." If, after the first minute, the person is just staring at the machine for four minutes, they could spend one minute at the next machine and then the next, so of course they could handle five machines.

But if the worker starts the machine, sits down and smokes a cigarette like they do in other countries, they will say that they are being made to work harder. The Japanese, on the other hand, don't even take time to smoke a cigarette. They don't think, "The machine is running now, so I'll take a break." Instead, they do things that don't even matter. If they are going to use their valuable energy anyway, they should use it to do work.

Machine cycle time and manual cycle time are still mixed up all over the world.

CHAPTER 32

Operational Availability vs. Rate of Operation

There is one other thing that is confused in people's thinking, and although this is another bit of wordplay, they are *kadoritsu*[40] (可動率) and *kadoritsu*[41] (稼働率). Just as it is written, operational availability is the rate (率) that you can (可) run (動) the machine. If the machine is broken and cannot be operated, the operational availability is poor. Naturally, when the operational availability is poor, the rate of operation is also poor. Operational availability should be raised as much as possible, and you should make efforts to achieve 100 percent.

In terms of the rate of operation, there is no point in running the machine if you do not have work. You will not make money. The rate of operation is determined by the external factors of whether or not you have work, and it is foolish to keep the machines running, making parts that will not sell, just because there is a large depreciation rate that shows a loss.

Another way to put this is that operational availability requires good PM.[42] If you can keep machines from breaking down, your

---

[40] Operational availability. The word 可動率 is also read *bekidoritsu* in Japanese in order to avoid confusion.
[41] Rate of operation.
[42] Preventive maintenance.

machines will be in such a condition that they can be run whenever you need them. But if you cannot run your machines because you have no work, this has nothing to do with operational availability. If machines are unavailable because of losses due to changeovers, you should work on reducing changeover times, and this will increase the operational availability.

The rate of operation changes depending on the amount of work you have at the time. You may need to work overtime to run 120 percent and at other times you may run at a much lower rate to save energy, or even shut the machines off.

However, when people mix the two up and say that they must increase both their operational availability and rate of operations, this is similar to my earlier point on how people mix up man-hours and machining time. If the machines really need to be running but they are broken and cannot run, you should do thorough maintenance so that they do not break down. Keep machines in operable condition so you can run them whenever needed. It is a problem when you have a lot of work and want to run your machines fully, but they are broken down and you cannot run them.

People confuse operational availability and rate of operation, and say, "We did not make any money because our rate of operation was bad." You need to look at which *kadoritsu* was bad. I think this confusion is a result of people feeling that it is a loss to leave the machines idle when they are in operable condition.

The spinning business is bad these days, so manufacturers are operating on reduced hours. Compared to their busiest times, they have cut operating hours by nearly 50 percent. Recently they scrapped a large number of their machines because their industry is structurally depressed due to overcapacity in the market. I am opposed to this. It is important to have the machines available to run again when needed, but when they scrap the machines the government gives them money. Then they take this money and buy the latest machines, and when the machines are fully automatic they say, "We can run these machines 24 hours a day." They will overproduce again. This is what we call "scrap and build."

When there is real growth this is a good thing, and rather than letting bad machines hold you back, you should scrap them and install the most modern equipment. When business is good, you will not qualify for a depressed industry so the government will not give you money even if you scrap your machines. Rather than doing this, the correct thing to do is to reduce cost, increase productivity, and become more competitive internationally.

I tell Toyoda Boshoku also that we must reduce our cost to below that of developing countries. If we cannot reduce our cost, we must consider getting out of the business. If a large spinning company decided to get out, probably more companies would also decide to get out of the business. If there were half as many large companies in this industry, the remaining companies may be able to go back to full operations. Even at full operations, if the cost is higher than that of developing countries, we may not be competitive. If you run unmanned machines for three shifts you can increase productivity five-fold, but if you overproduce, this is no good and productivity will go back down.

As I often say, the companies that really pursue how to make just the quantity needed at a lower cost will survive until the end.

At Toyoda Boshoku they have none of the new modern machines. So I told them that if they could not think of ways to use these machines to achieve the same man-hours as the latest machines, their only option was to get out of the textile business.

However, even with a slight change in the exchange rate, all of the efforts at cost reduction can disappear in an instant. This makes textiles a very challenging industry.

Wages will also go up by 8 percent or 9 percent. Increasing productivity just to keep up with this is hard enough. Normally, the Productivity Standard Principles[43] used for negotiating wages only takes into account direct labor productivity. But the more that modern manufacturing progresses with automation, the more they add indirect

---

[43] The use of the Productivity Standard Principles (生産性基準原理) is promoted by the Nippon Keidanren (Japan Business Federation) as a basis to determine wage increases.

labor instead of direct labor. When this happens and the productivity of a limited number of workers is improved by 10 percent and this productivity is used to justify wage increases of 10 percent, there is a large increase for the majority of people who are not part of this productivity calculation. If the raise was less than the increase in productivity, I think it would be okay to give wage increases to the staff and personnel from indirect departments. This is a tremendous pitfall, and when officials talk about productivity, they think it should be a 10 percent increase in productivity for the direct labor in spinning, if we take spinning as an example. This is true in all companies, not just spinning. But to increase the productivity of everyone in the company is impossible. It is hard enough to increase the productivity of direct labor by 10 percent, so to raise the productivity of the thousands or tens of thousands of other employees just can't be done.

CHAPTER 33

The Difference between Production Engineering and Manufacturing Engineering

We think of production engineering and manufacturing engineering as distinct things. We distinguish manufacturing engineering as the work to determine the method of manufacturing and production engineering as how to actually accomplish that method of manufacturing.

There is an old Japanese expression: "Fools and scissors are useful if handled properly." Taking an example that is close at hand, manufacturing engineering studies how to use scissors effectively to cut things well, while production engineering studies what is the right type of scissors for the job, such as a fabric scissors for cutting textiles or a pruning scissors for cutting tree branches. The work of production engineering also includes developing new scissors to cut a certain material. But without manufacturing engineering you will not be able to use these scissors properly.

Even if an inexperienced person borrows a pair of sheet metal shears, they cannot cut sheet metal with them. A person who is skilled at using sheet metal shears can also cut thin paper with them. In extreme cases, he may even use shears to give himself a shave.

As in the earlier example with the cameras, I believe that getting good results is not just about having good equipment. You cannot do good work unless you have learned how to use the equipment properly. Manufacturing engineering must be thoroughly studied, but just as in the example of rate of operation or processing time, people are confused about this.

Now that computers have become more common, people talk about hardware and software, but even before that we have been thinking seriously about the difference between manufacturing and production. We realized that if we had strong manufacturing engineering, we could do more and more *kaizen*.

We used to call manufacturing engineering "*gemba* engineering."

When I took my first trip to the United States more than 30 years ago, the person who toured me through the factory was a plant engineer. On his business card it said "General Plant Engineer," and he was very knowledgeable about the workings of the *gemba*. When I asked "What is that?" he would go find a foreman right away to ask, and it was clear that he communicated very well with the shop floor foreman. I returned to Japan feeling "We need to have engineers like him." We think of "*gemba* engineering" to be the plant engineer. Today, there is something called "plant engineering," but from what I understand this is not engineering on the shop floor.

It seems like plant engineering in Japan includes layout engineers also, but it would be better if they were more focused on shop floor engineering.

One person from another company said, "Toyota is fortunate to have good people as manufacturing engineers. The parent factory that oversees my factory has many good production engineers but, sadly, no manufacturing engineers." At that time I realized that what we were calling *gemba* engineering was seen as manufacturing engineering, and we began to use the name manufacturing engineering after that. That is why we would like to think of manufacturing engineering as plant engineering, but...

It is common for production engineers to become separated from the *gemba*. There are some professional entertainers whose act is to cut paper with scissors. These entertainers become separated from the scissors makers. The scissors maker will research not only the scissors used to cut paper. They also study scissors for cutting fabrics or silk. In the past there may have been one scissors used for everything. But sometimes clippers work well. Or fingernail clippers may be the right tool. It is important to continuously develop the right tools for the job.

Years ago I was disliked for saying to people in the Toyota Motors production engineering department, "You're just catalog engineers." They would look at a catalog and say "This is a good machine. This machine will greatly increase productivity, so please buy it." But this is not good at all, and I also said to them, "You haven't even developed any machines on your own."

Within a company, production engineering and manufacturing engineering really ought to be a single body.

When developing new products, or when new materials will be used, it is better if the production engineers evaluate the proper equipment and the processes required to make the product. Once the manufacturing engineers can use them properly, the work of manufacturing engineering should be to do process *kaizen* and equipment *kaizen*. This is a most important thing, but people who don't get it just don't get it.

CHAPTER 34

The Pitfall of Cost Calculation

Whenever we need to make a decision, we end up doing something like cost calculation. It is not wrong to do cost calculation, but I think top management sometimes makes the wrong judgment because of it. When the top management is told that according to the calculation the investment will pay off or the cost will be reduced, they may see that the numbers look right and agree to buy a machine. But after the machine is purchased, it is possible that they only have orders for 5,000 pieces and not 10,000 pieces as planned and the investment may never pay off. You can calculate the preconditions for making the investment pay off, such as what level of demand is required, what is required for a two-year payback, and so forth. However, when you do not sell as many as you had planned, this becomes a very expensive investment. When people want a certain machine badly, they will tend to do the calculations that suit them. Even if you complain later that the demand forecast was inaccurate, no one will make reparations for your losses.

There is another problem. This is another case where you can only really tell which is better based on the results, but there is something called the cost calculation method. Cost calculation includes the depreciation of the machine. If the depreciation schedule for a particular machine is ten years, then after ten years when the machine has been completely depreciated, they will scrap it and build or acquire the latest equipment. The real purpose of including depreciation as part of the cost calculation method is to allow for retained profit. After all of the years of use and retained profit, if you scrap the machine just because it has been completely depreciated, this is a disaster. In fact, you have not retained the profit at all.

Production engineers at profitable companies and at large companies will say, "That machine is fully depreciated. It is old and wearing out. Efficiency will go up even more if we buy a new machine." They fall into the misconception that they can use the money from the retained profit to buy the new machine.

Somehow the financial people cannot seem to understand that once the machine is fully depreciated and becomes retained profit, this machine can now actually be used at no cost. They do not understand that once a machine is fully depreciated, you can really start making money with it. When production volumes are growing and the economy is growing rapidly they say we should invest as much as we can in equipment, within what depreciation will allow. But when we take this thinking to times like we have today, when overproduction is a grave danger, we have a big problem. Rather than that, we can increase our profits if we produce at a low cost using the fully depreciated machines, even if the sales volumes do not grow. Instead, companies throw away the machines to save on taxes, buy new machines, and complain about not making money.

For general equipment, if we use "general-purpose dedicated machines" or "general-purpose automatic machines," then we can modify these machines and use them again, even with the product changes. This is how you can use your "free" machines—in other words, the machines that have been fully depreciated—to make money even when your products change.

The term "general-purpose dedicated machine" is very frustrating for designers. How can something be both general-purpose and dedicated? But I do see that there are machine tool manufacturers who are using these words in newspaper advertisements these days. It was immediately after the first oil shock when I asked for general-purpose dedicated machines to be built.

Depending on the industry, and this is true for Toyota and auto body manufacturers as well, the majority of the cost of equipment investment is the cost of making dies. Even at Toyota 60 percent of the equipment investment cost is in the dies.

CHAPTER 35

The *Monaka* System

Dies have no versatility. When a new product is introduced and the previous one is discontinued, we have to throw away the old tooling and make new tooling. That is why press dies are not general-purpose items.

When the body style of an automobile changes, the body panel manufacturers have to make hundreds of large, all-new dies. Although these items can be depreciated, there is no retained profit, so this is quite a cost for manufacturers. However, the presses themselves are versatile machines so we had to think of ways to make our dies versatile. This is how we decided to make "sweet bean *monaka*"[44] type dies.

The bread-like shell of the *monaka* does not go bad easily, so you can make them ahead of time. When there is a school sports day somewhere tomorrow and they order several hundred *monakas*, you just need to cook the sweet bean filling the night before. You can make up the shells during your spare time. When the sweet bean filling is ready, you can add it to the shell and sell them.

By contrast, if you don't even make the shells because you have no orders and you sit around every day doing no work, you will not

---

[44] *Monaka* is a Japanese sweet made of a thin bread-like crust and a sweet adzuki bean filling.

be able to keep up with making the shells and cooking sweet bean paste when you get a big order because of a Buddhist memorial service.

Likewise, with the dies in a press, the "shell" can be used for anything. If the core of the die is hollowed out, and you build and change out the "sweet bean" section, you can make dies rather quickly. We are actually trying this concept in one area of our company.

But dies have traditionally been made as a single unit. I said they should limit it to three types for the outer part of the die: large, medium, and small. Then we should hollow out the core of these dies and keep them safely in storage. Then, when an old model is needed again we can build it inexpensively by adding the old die into the outer part. But this is fairly difficult to do, and it will take more than five or ten years to make all of our dies like this.

We thought if the *monaka* system worked, the "sweet bean rollup" method might also be good so we tried that, but the sweet bean rollup method did not work. The sweet bean rollup is made by preparing the sweet bean filling ahead of time and then pouring the flour onto a metal griddle and rolling up the sweet bean paste. So this is the reverse of the *monaka* system. The sweet bean goes bad sooner and costs the most. The dough on the outside is cheap. If you make a large amount of sweet bean filling ahead of time without even knowing when you will sell it, and then you don't have any customers, the sweet bean filling will go bad.

On the other hand, you have to make the new dies by a certain deadline. So we set the standards for the outer dimensions ahead of time. When the die is no longer needed, we can hollow out the center. At about the 30,000th unit of production, it becomes time to make the filling. But that is what takes most of the time.

In the future, we can reduce the equipment investment for dies by giving the die itself versatility while making just the filling a dedicated core.

As long as you make the dies slightly larger than you need, you can still use them even if the new parts are bigger. On the other hand, if you make dies that are dedicated to a certain part, making them just big enough, thinking you are "trimming the fat," then you will no longer be able to use the dies even when the new parts are slightly larger. This type of versatility is important. The ability to develop these sorts of things one after the next is a characteristic of the Toyota System. It is often said that running a business relies on three elements—in other words, people, materials, and money. But in the end, unless a company generates a profit, it cannot fulfill its social responsibility. One way a company can generate a profit is by how they sell—in other words, to make money by trading. A company can also generate a profit by the wise use of money. And the other way is to reduce cost, and these are basically the three ways for a company to make money.

As long as you can make money through trade, it is easiest, since whether you lower the cost or whether the cost is kept the same but you raise the price, you can make a profit.

Perhaps when the person at the head of the company is good at trading, and they are able to bring in higher and higher profits, the *gemba* can take it easy and still get by. But when it becomes harder to succeed at trading, and when the financial markets are tighter, it is not as though you can do drastic cost reduction. So after all, cost reduction is something that should be done from the beginning, as the first thing. What I want to say is that this is the most important work of engineers, or shall I say the *gemba*.

What is the reason that we reduce inventory? It is to make the finances easier.

For example, what if 500,000,000 yen of work in process and inventory was reduced and this 500,000,000 yen was in the accountant's safe? The accountant could invest this money in some sort of market- able securities, and we could make a profit of several percent from this 500,000,000 yen. But when this money is on the *gemba* in the form of materials, we now have to borrow money. We have to pay the material supplier for the materials, and pay the electric company for electricity. So you see, the difference between profit and loss here can be very large.

If you managed your inventory well you could pay out dividends, but instead we have to pay the bank 10 percent and we can only pay out 5 percent to the stockholders, and this is a very foolish way to operate. In the end, if the *gemba* reduced inventory instead of doing silly things, and if that cash was in the accountant's safe, the difference would be huge. The dividends could possibly double, and if our profit increases and we pay a lot of taxes, then this will also benefit the country. Instead, when people hear "cost reduction" they tend to think of it as the role of accounting. Accounting cannot do any cost reduction.

CHAPTER 36

Only the *Gemba* Can Do Cost Reduction

I was telling the person in charge of personnel the other day, "When the *gemba* says they want 100 people, you should give them 10 people. Then the *gemba* will find one way or another to meet their production requirements. So if they come crying to you that they can't make it unless they get 100 people, just give them about 10 people and then ignore them. Then even human resources can achieve a cost reduction of 90 people."

The same is true of accounting. The shop floor reduces inventory. This money goes into the bank. This cash can be used to generate a few percentage points of profit. So even accounting can reduce cost if they effectively use the money that the shop floor saves. Instead, accounting thinks it just needs to allocate cost savings targets. Even if accounting identifies the cost structure needed to be profitable, or asks the manufacturing department to reduce cost by several percent because they are losing money, or asks the design department to reduce cost by several percent, these allocations are useless unless the others actually do the cost reduction.

Therefore, the *gemba* must become fanatical about cost reduction, with the belief that only the *gemba* can do cost reduction. People are overly concerned with cost knowledge, but this displaces cost consciousness. I say you don't need cost knowledge. I'm not even interested in learning the terminology.

If you use calculations based on cost knowledge, you can show that the cost has been reduced or increased. When you use your consciousness to think, the answer is very clear. But when people use calculations to show that equipment investment will pay off, or that the cost will be reduced, I say this is foolishness.

It is really bad math to say that you will achieve rationalization goals by such and such percent each month. Some months you may see no improvement at all. When volumes increase and you are able to produce more without adding machines or people, you may see a big improvement. This is why the accumulation of your *kaizen* bit by bit may show results much later. But if you give up on this and implement computers or robots, you will not be able to use them right away. You must not give up on the accumulation of your daily efforts.

Between 1955 and either 1972 or 1973 we could sell as much as we could produce. During times like those we could see the results of rationalization quickly, as we were able to avoid adding people and the rate of operation of our machines improved, which reduced cost. That is why the impact of rationalization was so big.

On the other hand, when the economy slows down, you really must have perseverance. When you need to validate your results or try to meet your calculated targets, you may begin to panic and wonder, "How are we doing this month? What about next month?" and this is absolutely the wrong thing to do. The trouble is, perseverance is hard.

In these days people use the word "perseverance" more often. The Japanese professional golfer Aoki is known as a man of perseverance. Even sports become a game of perseverance, once everyone is at a comparable level of skill.

As long as there is a gap in the skill level between athletes, we can have "undisputed leaders" or "the great ones," but as the gap in skill or technique gets smaller, perseverance makes the difference.

Even though he frequently plays with perseverance, Aoki still loses sometimes. But if you lose and then panic, you become like the golfer Ozaki.

If the UAW at General Motors had the perseverance to accept a reduction in wages, they would become a formidable competitor to Toyota. In Japan today, suggesting a reduction in wages would cause an uproar....

CHAPTER 37

Follow the Decisions
That Were Made

It is important to "follow the decisions that were made" as a fundamental way of working. Whether it is implementing the *kanban* system or implementing anything, it is difficult to follow the rules that have been set. It is easy to say these words, but not easy to do. So, why can't we do what we decided to do?

Apparently, the words "decisions that were made" have a funny ring to young people's ears. They think their superiors made the decisions. Why don't you decide? I say. That is *kaizen*. First, try out the rules that were set. Then, if you cannot follow the rules, you should think there is something wrong with the rules.

If we take *kanban*, for example, if you cannot follow the rules that have been set, that just means there is something wrong with the way the *kanban* system was set up. When someone gives an opinion that it would work better by doing it another way, you should immediately try it. That way, the decisions that were made become the rules made by you, for yourself.

It does not work when people think of "the decisions that were made" as decisions made from above. There was a lot of that at the *gemba* in the old days. When a new worker came in and gave his ideas for making things better, the foremen and workers would say, "He doesn't know his place," and would not take up the new ideas.

All of these sorts of improvements, thinking "This is how I will change it," and making the change, is *kaizen*. But sometimes this can be *kaiaku*.[45] If it is a change for the worse, just fix it right away. Don't change it back to the way it was, but try something else.

There is something called standard work,[46] but standards should be changing constantly. Instead, if you think of the standard as the best you can do, it's all over. The standard is only a baseline for doing further *kaizen*. It is *kaiaku* if things get worse than now, and it is *kaizen* if things get better than now. Standards are set arbitrarily by humans, so how can they not change?

When creating standard work, it will be difficult to establish a standard if you are trying to achieve "the best way." This is a big mistake. Document exactly what you are doing now. If you make it better than now, it is *kaizen*. If not, and you establish the best possible way, the motivation for *kaizen* will be gone. That is why one way of motivating people to do *kaizen* is to create a poor standard. But don't make it too bad. Without some standard, you can't say, "We made it better," because there is nothing to compare it to, so you must create a standard for comparison. Take that standard, and if the work is not easy to perform, give many suggestions and do *kaizen*.

We need to use the words "you made," as in "follow the decisions you made." When we say "that were made," people feel like it was forced upon them. When a decision is made, we need to ask who made the decision. Since you also have the authority to decide, if you decide, you must at least follow your decision, and then this will not be a decision forced upon you at all.

But in the beginning, you must perform the standard work, and as you do, you should find things you don't like, and you will think of

---

[45] Change for the worse. *Kaiaku* (改悪) is the opposite of *kaizen*.

[46] Standard work (標準作業) is called *standardized work* at Toyota today and is not the same as work standards. Standard work is defined as "the most effective combination of manpower, material, and machinery" and is built on three elements, *takt* time (pace of customer demand), work sequence, and standard work in process.

one *kaizen* idea after another. Then you should implement these ideas right away and make this the new standard.

Years ago, I made them hang the standard work documents[47] on the shop floor. After a year I said to a team leader, "The color of the paper has changed, which means you have been doing it the same way, so you have been a salary thief for the last year." I said, "What do you come to work to do each day? If you are observing every day you ought to be finding things you don't like and rewriting the standard immediately. Even if the document hanging here is from last month, this is wrong." At Toyota in the beginning we had the team leaders write down the dates on the standard work sheets when they hung them. This gave me a good reason to scold the team leaders, saying, "Have you been goofing off all month?"

If it takes one or two months to create these documents, this is nonsense. You should not create these away from the job. See what is happening on the *gemba* and write it down.

---

[47] Standard work documents refer to the standard work sheet showing the layout of the line or cell, people, work flow, standard work in process placement, quality, and safety checkpoints. The standard work combination sheet graphs of the manual work and automatic time for each individual against *takt* time (pace of customer demand). The two are sometimes combined to make the standard work instruction sheet.

CHAPTER 38

The Standard Time Should Be the Shortest Time

Speaking of standards, time study is another thing everyone gets wrong. For example, people measure ten repetitions of a task and use the average value. I think this is the worst thing you could do. If you are watching a person doing something ten times, and if they are doing it differently each time, you should immediately correct them. Instead, people think, "That's not my concern. I just learned the symbols and how to use a stopwatch and I write things down. And after measuring ten repetitions, the standard time will be set as the average of the ten times." If you are going to take ten such unreliable measurements, you should choose the shortest time. Some say that is harsh, but what is harsh about this? The shortest time is the easiest method.

Even if the ten repetitions are performed in ten similar ways, they are doing different things. The shortest time out of these is the easiest time. Therefore, you need to analyze why the others took several extra minutes or several extra seconds. Some say, "You can't always do the work the same way," but the times are different because something is wrong. I say there is nothing harder than to do the work in the average time.

Drop a nut once and pick it up. Working at the average time is like trying to catch the nut halfway because letting it drop all the way down takes too long. Who could do such difficult work? You can do the work in the shortest time if you ask, "Why did I grip the nut in a way that made me drop it?" and, "Is there a method to grip the nut more securely?" This will also be the easiest motion. Or you may find that people are pacing themselves because if they do a lot of work their workloads for the day will increase.

They say you should give a time allowance of a few percent for biological needs when you are setting standard times, but I say it should be zero. In the end, the difference between reality and what is measured is too great for this also. People include allowance times and changeover times based on trumped-up reasoning. This is where managers can be very tricky.

If a person needs to urinate, they should stop the line and go. If they need to take a break, let them take a break. However, they should summon a team leader and ask them to take their place while they go urinate. Even if the worker is feeling a bit ill and has to urinate more often, it is better to have them come to work and go to the bathroom three or four times during the morning than to have them stay home.

It is unacceptable to create slack time by determining the average and setting the rule that people can urinate every two hours during the day, and say it is every two hours even if a person is not feeling well, or that a person should go every two hours even if they do not need to. This is because everyone thinks in terms of average values, but there is no such thing as an average value in this world.

This is a silly story, but years ago when I would go to the machine shop and stand and watch people working for a while, I would see them going to sharpen their cutting tools again and again. While they were being watched, they were making it so we could not measure them. The person doing the measurement thinks, "They don't sharpen their tools that often." So they don't say anything.

One person who went to measure a woodworker's time said that the woodworker kept sharpening his plane and would not let him take time measurements. The woodworker would take two or three passes at sharpening his plane, then shake his head and sharpen it some more. That is why it is totally meaningless for a person who cannot do the work to measure the time.

In the end, if you are going to take several measurements in order to set the standard, you should take the shortest time. Then, it is important to find out why people cannot do the work within this time and to teach them in a way that they will be able to do the work within this time.

AFTERWORD

Publisher's Note: The original text contained no afterword. The following is a passage from Taiichi Ohno's introduction to the first textbook on the Toyota Production System (TPS), created in 1973 by the Education Department at Toyota Motors:

> "Whatever name you may give our system, there are parts of it that are so far removed from generally accepted ideas (common sense) that if you do it only halfway, it can actually make things worse.
>
> "If you are going to do TPS you must do it all the way. You also need to change the way you think. You need to change how you look at things.
>
> "Just as magicians have their tricks, *gemba* engineering has its tricks. The magician's trick in this case is 'the relentless elimination of waste.' In order to eliminate waste, you must develop eyes to see waste, and think of how you can eliminate the wastes that you see. And we must repeat this process.
>
> "Forever and ever, neither tiring nor ceasing."
>
> *Taiichi Ohno*

ABOUT THE AUTHOR

Taiichi Ohno was born in Dalian, China, on February 29, 1912. He joined Toyoda Boshoku in 1932 after graduating from the mechanical engineering department of Nagoya Technical High School. He was transferred to Toyota Motor Company in 1943, and he was named the machine shop manager in 1949. Mr. Ohno was promoted at Toyota to director in 1954, managing director in 1964, senior managing director in 1970, and executive vice president in 1975. He retired from Toyota in 1978.

Mr. Ohno is the father of the Toyota Production System. He authored three works: *Toyota Production System: Beyond Large-Scale Production*, *Taiichi Ohno's Workplace Management*, and *Just-in-Time for Today and Tomorrow* with Setsuo Mito.

Taiichi Ohno died on May 28, 1990, in Toyota City, Japan.

SEEKING WHAT
TAIICHI OHNO SOUGHT

Today the world is increasingly aware of the power of the Toyota Way as a winning business philosophy and management system. While there is growing understanding of its systems and tools, there is still much to be learned about the core values and guiding principles that make Toyota the most successful manufacturer and one of the most respected businesses in history.

Taiichi Ohno was the father of the Toyota Production System, but more importantly his teachings continue to influence the thinking of the leaders at Toyota today. This book offers a glimpse into one of the greatest minds in modern management. Some of the ideas you will find here may surprise you, and can change both how you work and how you look at work itself. Ohno seeks to overturn conventions, destroy misconceptions, and go beyond common sense.

There are a number of things we must keep in mind when reading *Taiichi Ohno's Workplace Management*. First, this text is not a book written by Taiichi Ohno. It is the spoken narrative of his ideas on management and the experiences that shaped those ideas. This text comes from a series of interviews conducted by the staff of JMA (Japan Management Association) in 1982, and published later that year. As such, readers are encouraged to read as though Taiichi Ohno is speaking directly to them, rather than to read looking for a logical theoretical structure to the text.

Second, if you are new to *kaizen*, the Toyota Production System, or so-called lean manufacturing, this book ought not to be the very first book you read on these topics. While the author mentions concepts such as *kanban*, *kaizen*, and *heijunka*, he does not always explain them in depth. This book is not an introductory text or a "how-to" for implementing the Toyota Production System, but it can serve as a resource to provide context to and deepen your understanding of the Toyota Way once you are familiar with the basics.

Third, we must heed the words of haiku poet Matsu Basho (1644–1694), who wrote:

"Do not seek to follow in the footsteps of the old masters, seek instead what these masters sought."

「古人の跡を求めず 古人の求めたるところを求めよ」松尾芭蕉

Many people today are seeking to build their own winning *gemba* management system, just like the one built by Taiichi Ohno at Toyota. The majority of seekers do not come from industrial companies. The study and application of *kaizen* and the Toyota Production System has become increasingly a part of how hospitals, governments, universities, banks, mining operations, and retailers are choosing to improve performance and develop their people. As Taiichi Ohno's legacy expands far beyond manufacturing, it is increasingly important that we do not simply follow in the footsteps of old masters.

So, what was Taiichi Ohno seeking? To answer this question we need to pay careful attention to his ideals, his thinking style, his approach to learning and teaching, his views on frontline leadership, to his spirit, and his character. We must seek not to imitate the what, the great exploits of masters such as Taiichi Ohno. We must study and strive to understand how and why they arrived at their accomplishments. Only then can we leave footprints on our own paths.

In more than one instance, Taiichi Ohno prefaced his words with the invitation, "I heartily welcome criticism and comments from our readers." Therein lies the first hint into the master's character.

Jon Miller, CEO
Kaizen Institute
Seattle

OHNO'S INSIGHTS
ON HUMAN NATURE

Taiichi Ohno, credited as the architect of Toyota's Production System (TPS), was concerned that human nature would stand in the way of managers' ability to understand TPS and achieve continuous flow. In his 1988 book, *Toyota Production System* (Productivity Press), Ohno noted that people are accustomed to processing work using the batch-and-queue method (p. 10), which means they like to stockpile raw materials, work in process, and finished goods. He said inventories reflect a natural human behavior to hoard things in preparation for bad times, but that we should not get stuck on this way of thinking because it is no longer practical in demand-driven buyers' markets (pp. 14–15). Ohno said it would require a "revolution in consciousness" by business people to overcome their obsession for hoarding. Indeed.

While Ohno was no doubt correct, most business leaders do not like revolutions of any kind. Frederick Winslow Taylor said in the early 1900s that his Scientific Management system required a "mental revolution" by managers, particularly with respect to improving relationships between management and labor. Most business leaders did not like Taylor's "mental revolution" idea. They were far more comfortable with evolution than revolution. But did they actually evolve when it came to making fundamental process improvements, as Taylor suggested? Some did, but most did not. And most have not since then.

Chapter 1 of *Workplace Management*, "The Wise Mend Their Ways," describes the need for leaders to avoid mental rigidity, to not fear change, and to be humble as prerequisites for adapting to change.

Since batch-and-queue thinking is so deeply embedded in the human brain, extraordinary effort must be applied to break free of this way of thinking. Normally, when we possess a physical or mental habit that we want to change, we commit ourselves to the daily practice of new routines. If we want to break free of batch-and-queue thinking, we have to learn to see batches and queues of material and information and engage in new and unfamiliar concepts and processes to reduce or eliminate them. If leaders cannot do this, then they cannot adapt.

After many years of effort, the best that most organizations have been able to do is process material and information in a hybrid batch-and-queue/flow way. This outcome illustrates how challenging it is for leaders to adapt. Continuous flow remains elusive, which underscores Ohno's point that our basic nature is to hoard things in preparation for bad times.

In recent years, attempts to achieve flow have been disrupted by managers who view decoupling of processes in a value stream as a more efficient and lower-cost way to process material and information. Different types of work that were at one time done in close proximity are now distributed across the globe. Design work is done in California; engineering is done in Connecticut; manufacturing is done in China; assembly is done in Mexico; and customer service is done in India. Human nature does indeed stand in the way of understanding and achieving flow.

As an educator, my fundamental objective is to teach managers that leading organizations for flow is different, both broadly and in detail, from leading organizations for batch-and-queue (or hybrid). Ohno understood that managers' beliefs, behaviors, and competencies are completely different. And he clearly understood that to get good at anything you have to understand the details. This is where most managers fall down with respect to TPS; they do not want to understand the details, and therefore fail to adapt.

Chapter 2 of *Workplace Management*, "If You Are Wrong, Admit It," describes illusions in leaders' thinking that prevent them from trying new things and making mistakes, which, in turn, helps them avoid having to admit they are wrong. For many years I have taught managers how to lead TPS and have learned quite a few interesting things, some of which are listed next. They reflect an aversion by managers to making mistakes (or the possibility thereof), admitting they are wrong, or conceding that they do not know or understand something—especially to subordinates.

▲ If details about leading TPS are provided in ways that are easy to understand, some managers will say, "I already know that." But they surely do not; they think they know it because it has been presented in an easy-to-understand way. They confuse knowing with doing. They do not know or do flow.

▲ If details about leading TPS are provided in ways that are challenging to understand, some managers will say, "That's too much detail." The details, of course, are critical to TPS success. Ask any professional musician, golfer, visual artist, opera singer, etc., about the importance of details.

Chapter 4 of *Workplace Management* says, "Confirm Failures with Your Own Eyes." Managers who are serious about improvement are not afraid to experiment and observe what happens.

▲ If insufficient details are provided, then some managers will say, "That's too high-level. Give me more specifics." However, the specifics are what managers must learn through their own daily application of Toyota Way principles and TPS practices. I cannot do it for them. An old and renowned piano teacher once told his most accomplished young student, "Now you must make the piano sing." Just that; nothing more specific. For serious students, that advice is more than sufficient for them discover the next level of detail. Similarly, Ohno would say to his most accomplished students, "You must think for yourself." Just that; nothing more specific.

▲ If insufficient details are provided, then some managers will say, "That's theory." Managers misuse the word theory to describe something that they are not familiar with. Theory, of course, is an explanation for an experimentally testable hypothesis that others can replicate via experiments. TPS is not theory.

As Ohno says in Chapter 5, "…misconceptions easily turn into common sense." Avoid stasis by going beyond common sense and trying new things.

▲ Even if I can absolutely, unquestionably, irrefutably, categorically, infallibly, and conclusively prove, with God in total agreement, that leading TPS will do great things for an organization, most managers will say, "No thanks." They will get the business to where it needs to be by other means. They do not want to understand the details. They do not want to let go of their batch-and-queue (hoarding) mentality.

Chapter 21 of *Workplace Management* says, "'Rationalization' Is to Do What Is Rational." Leaders often refuse to break down their misconceptions and do what is rational.

Ohno taught us that TPS is a completely different way of thinking and doing things. He also taught us that TPS must be led by managers because if managers do not lead, then TPS will have no chance as an overall management system. Instead, it will instead exist ephemerally as an assortment of tools, often used incorrectly, to cut costs and improve productivity (mainly in operations). And it will invariably result in bad outcome for workers.

This is what we see in most organizations. It is the popular version of TPS—fake TPS—which is formulaic and incomplete, and does

nothing to challenge the hoarding status quo. Ohno would be shocked to see how abundant fake TPS is today, and disappointed with the performance of so many people in leadership positions. Their inability to try new things and make mistakes, admit they are wrong, or concede that they do not know or understand something means they are unwilling to "change the way [they] think" and do TPS "all the way" (see Afterword).

Professor Bob Emiliani
Connecticut State University
School of Engineering and Technology
New Britain, Connecticut

A REVOLUTION
IN CONSCIOUSNESS

————————

Much has happened in the world of manufacturing since the birth of Taiichi Ohno 100 years ago. His revolutionary Toyota Production System has disseminated far and wide, especially since the seminal book *The Machine That Changed the World* (Harper Perennial, 1991) introduced it to the world as "lean production." Exactly how much has happened, and how we might evaluate that, depends on how we define exactly what "lean" is.

Also 100 years ago, Henry Ford was preparing to open his historic Highland Park assembly plant where he would show the power of flow production—central to what we now call lean thinking—to the world. Around the same time numerous innovations essential to lean thinking emerged: the birth of industrial engineering as the science of efficient work design, the genesis of modern psychology as a true science sowing the seeds of today's discoveries in the neuroscience of human learning as well as modern theories of organizational learning, and the framing of the scientific method as it forms the basis of practical lean problem solving. *Lean* isn't lean without all of these.

One hundred years ago Toyota the global auto giant was still Toyoda the small but ambitious Japanese loom company. Group founder Sakichi Toyoda was hard at work developing the automatic loom that would provide the funding for his son Kiichiro to launch their auto business. Ohno was born in China, where Sakichi would soon set up his ultimate loom factory and perfect his loom, and where Ford's motorization dream reached its ultimate apex (Chinese auto makers sold over 18 million vehicles last year, more than was ever sold in the United States even at its peak, and the Chinese market is still maturing).

But Ohno's revolution wasn't limited to auto production. His deep message is that even the best of business systems are a process-in-process and continuous improvement requires a "revolution in consciousness." The Toyota Production System was developed not by grand design but by emergent problem-solving and experimentation. As Ohno states in another collection of his wisdom *The Birth of Lean* (Lean Enterprise Institute, Inc., 2009): "We are doomed to failure without a daily destruction of our various preconceptions."

Much has happened in 100 years, but how much progress? What would Henry Ford think of his dream of motorizing the world? What would Taiichi Ohno think if he could see how his production system has proliferated, propagated, and disseminated? I suspect they would both have mixed feelings. Dissemination? Yes. But what about propagation of the true intent?

I imagine both Ford and Ohno would be appalled with much of the current state of industry. Ford would be pleased to see the incredible turnaround of the company that still bears his name. But he would surely be baffled and disgusted with the modern-management corporation with its layers of conference-room managers, rigid organization charts, and impediments to continuous experimentation. (No doubt he would also question why people think they need so many superfluous bits of technology on their cars when his Model T was, actually, just fine.)

As for Ohno, surely he would be astonished at the incredibly wide-ranging dissemination of his ideas. Just as surely, he would be distressed by the all-too-common focus of many practitioners to apply the various lean tools without linking them to deeper purpose.

I'd like to think that Ohno would celebrate his centennial by drawing a fresh "Ohno Circle" (where he would identify a good spot to observe the frontline, real value-creating work of the business), and observe the way work is done in 2012 to find deep, even revolutionary, improvements. In that spirit, that's exactly what we should all do, pressing forward to new frontiers while continuing to deepen the fundamentals, asking ourselves: what preconceptions shall I destroy today? As Ohno says in *The Birth of Lean*: "If you're going to do *kaizen* continuously, you've got to assume that things are a mess." I hope this new edition of Ohno's deeply insightful *Workplace Management* will inspire us all to do exactly that.

John Shook
Chairman and CEO
Lean Enterprise Institute

TAIICHI OHNO
AS MASTER TRAINER

Many people know Taiichi Ohno as the lead developer of the Toyota Production System, but I believe his larger contribution was in the development of people. In fact, many people were involved in developing TPS inside the company, and some like Shigeo Shingo from outside. What continued its development as a system were the many students of Ohno, and the students of students. Many of the most senior executives at Toyota were Ohno trained; the best known was Fujio Cho, the past president and at this time chairman of Toyota.

There is no question that Ohno was a manufacturing genius, envisioning possibilities that few others could imagine. But as he learned the vagaries of TPS implementation, it became clear that the strength was in *kaizen* done by the people at the *gemba*. Continuous improvement depended on skilled and passionate leadership that would not let a day go by without *kaizen*. Ohno developed many methods to train them, but in reflecting on stories I have heard from students, it came down to a few principles:

1. *Learn at the* gemba. Ohno did not believe in classroom training. Training was hands-on.
2. *The teacher must stay ahead of the student in learning.* Ohno himself was an obsessive learner, always at the *gemba* improving TPS and improving himself, and he saw no end point for learning.

3. *Be a tough coach with high standards.* Ohno was the tough, demanding coach who would not allow the student to relax or settle for anything less than perfection.
4. *Love your students.* I recall an interview I did with Mr. Cho where he described how demanding Ohno could be, rarely giving a compliment, but at the end of the day he would gather his students around, and it was clear he was working so hard to help his students develop themselves because he loved them all. Cho had a tear in his eye when telling the story.
5. *Always be passionate, even obsessed with* kaizen. Your passion will rub off on your students, as will any lack of passion.

Ohno was following in the footsteps of great master teachers who had apprentices and taught complex skills, for example, arts, music, trades, martial arts, and cooking. There are now many books by master black belts from the martial arts who describe their methods of teaching "personal mastery," descriptions that sound strikingly like Ohno's approach.

As an example, I met an Ohno student from a second-tier Toyota supplier. He had been sent to Michigan as the manager of a plant that served Denso in order to raise the level of TPS in the plant. He described how Ohno's teaching "changed my life." I asked him what he meant by that. He said as a young industrial engineering student he had learned to set up jobs to achieve 85 percent work so that there was ample time for rest in each work cycle. Ohno taught him the only acceptable goal was 100 percent. Pursuing 100 percent changed his entire approach to industrial engineering and ultimately to leadership and life. Now I have told this story many times, and a typical reaction is that if the job is loaded to 100 percent the person will be overworked, as he or she cannot do value-added work 100 percent of the time. I explain that Ohno also preached that there is always waste in every work process, so you will never actually achieve 100 percent. But if you aim for 85 percent that is the *best* you can ever do. In reality if you measure yourself at 85 percent the worker is probably loaded to less than 65 percent value-added work because there is always wasted effort, and the worker will learn to do the job in less time then it initially took when it was set up. *Kaizen* is the pursuit of perfection, not the pursuit of good enough.

Now that the Toyota Way has been documented by Toyota, you can read about the core values—challenge, *kaizen*, respect, go and see, and teamwork. Ohno lived all of these values, but the passion for excellence always began with the spirit of challenge, and no challenge seemed impossible in the world Ohno built.

Jeffrey Liker
Professor, University of Michigan
Author of The Toyota Way

REFLECTIONS ON THE CENTENARY OF TAIICHI OHNO

Taiichi Ohno was born 100 years ago, on February 29, 1912, in Dalian, China, as the son of Ichizou Ohno, who was at that time an engineer of refractory bricks at the Manchurian Railway Company, a Japanese government arm for managing and developing Manchuria.

His first name "Taiichi" was taken from the "taika renga," which means refractory or fire-resistant bricks. The word "tai" means to endure and persevere. "Ichi" means number one. It also means to concentrate.

After returning to Kariya City in Japan, Ichizou set up a refractory bricks company and became its chief engineer. Later, he went into politics and became the mayor of Kariya City, a member of the prefectural congress, and finally a representative in the Japanese National Diet. While he was mayor, Ichizou assisted Toyota management in identifying locations for setting up new plants. Today, many Toyota group companies are located in Kariya.

In his middle and high school days, Taiichi Ohno was an active sportsman and a member of football teams. Perhaps this influenced him, as he was fond of saying later in life:

"Teaching means to teach something unknown. Training means to repeatedly practice something you know until your body remembers it."

Upon graduation from the engineering college in 1932, he joined Toyoda Spinning and Weaving Company, where he learned such practices as *jidoka* and multiple-machine handling developed by Sakichi Toyoda, the founder of the Toyota group of companies.

In 1943, he was transferred to Toyoda Motor Company when his company was absorbed by it. His experiences at his former company helped Ohno to develop Toyota Production System, which embraced many unique practices like *jidoka*, multiple-process handling, and continuous flow, which he had learned during those years.

The rest of Taiichi Ohno's career is now history. Today, he is remembered as a person who built a management system called such names as Toyota Production System, lean production, and Just in Time production, all over the world. Ohno changed the way to make products and increasingly how we deliver service in hospitals and even the public sector.

I had the unique privilege of spending time with Taiichi Ohno while accompanying him on his journeys to the United States, New Zealand, and Australia. This allowed me to stay close to the great man's "voices and coughs," as we say in Japan. I even played golf with him!

Once he asked me how the terms *kaizen* and *kairyo* (reform) were differentiated in the West. I said that while *kaizen* means to make improvement by using brains, *kairyo* means to make improvement by using money, and that in the West, most managers only think of improvement in terms of money. He liked this definition and quoted it on several occasions during his public speeches.

Although very few of today's business leaders have met or heard directly from Taiichi Ohno, the impact of his ideas and deeds is widely felt. He left an anthology of his sayings and axioms on management. I will mention a few of them in his memory.

> "Let the flow manage the processes, and not let management manage the flow."

In the lean approach, the starting point of the information flow is the final assembly process, or where the customer order is provided, and then the flow goes upstream by means of a pull signal such as *kanban*. On the other hand, the flow of materials moves downstream from the raw material stage to the final assembly. In both cases the flow should be maintained smoothly without interruption.

Unfortunately, in a majority of companies today, the flow is disrupted and meddled with by the convenience of the shop floor management.

"Machines do not break down; people cause them to break."

His life-long pursuit was to make a smooth and undisturbed flow as a foundation of all good operations. He believed that wherever and whenever the flow is disrupted, there is an opportunity to do *kaizen*.

"The *gemba* and the *gembutsu* have the information. We must listen to them."

Taiichi Ohno always placed respect for the worker first in his approach to *kaizen*. His focus was always on the customer, both external and internal.

"Just in Time means that customer delight is directly transmitted to those who are making the product."

Ohno was a man of deeds. Learning by doing was his motto and he did not engage in empty discussions. You pay money to buy books and go to seminars and gain new knowledge. But knowledge is knowledge, nothing more.

"Knowledge is something you buy with the money. Wisdom is something you acquire by doing it."

But you gain the wisdom only after you have done it. The real understanding of the lean operations is gained only after you have done it. No matter how many pages you may read on lean books, you know nothing if you have not done it.

"To understand means to be able to do."

It is with fondness and tremendous gratitude that I remember the great man Taiichi Ohno in this 100th year of his birth.

Masaaki Imai
Founder
Kaizen Institute
Tokyo, Japan

SELECTED SAYINGS
OF TAIICHI OHNO

On Teamwork

I used to tell production workers one of my favorite stories about a boat rowed by eight men. One rower might feel he is stronger than the next and row twice as hard. This extra effort upsets the boat's process and moves it off course.

On Standards

Where there is no standard, there can be no kaizen.

On the Seven Types of Waste

I don't know who came up with it but people often talk about "the seven types of waste." This might have started when the book came out, but waste is not limited to seven types. There's an old expression: "He without bad habits has seven," meaning even if you think there's no waste you will find at least seven types. So I came up with overproduction, waiting, etc., but that doesn't mean there are only seven types. So don't bother thinking about "what type of waste is this?" Just get on with it and do kaizen.

On Overproduction

If you asked me "What is the most important part of production control?" I would say it is to limit overproduction. If you can get away with staring at the floor until the scolding ends whenever "the parts were made" then production control is not doing its job at all.

If you put "at a lower cost" first you can make various mistakes such as overproducing or not making enough, or getting the timing wrong. There is no end to the pursuit of the Toyota System and how to produce at a lower cost.

On the Hoarding Instinct

We must not remain an agricultural people. We must become hunters and have the courage to acquire what we need, when we need it, in the amount we need. It goes beyond courage. I want this to become common sense in today's industrialized society.

On Kanban

The aim of kanban is to make troubles come to the surface and link them to kaizen activity. I tell people, "Let idle people play rather than do unnecessary work."

On How the Toyota Production System Came About

As a matter of fact, we could say that the Toyota Production System came about as a result of the sum of, and as the application of, the behavior by Toyota people to scientifically approach matters by asking "Why?" five times.

On Standardized Work

Standardized work at Toyota is a framework for kaizen improvements. We start by adopting some kind—any kind—

of work standards for a job. Then we tackle one improvement after another, trial and error.

On Practice over Theory

Don't look with your eyes, look with your feet. Don't think with you head, think with your hands.

On Walking the Gemba

It should take you hours to walk 100 meters each time you enter the factory. If it takes you no time at all to walk 100 meters that means no one is relying on you.

On Production Lines That Never Stop

The production line that never stops is either excellent or terrible.

On the Contradictions within Just in Time

To commonsense thinking it seems that Just in Time is full of contradictions, such as that between Just in Time and productivity, or between Just in Time and cost, or even the squeeze Just in Time puts on suppliers. We must break through this wall of common sense, and go "beyond common sense" in order to take the two contradictory sides and make them stand up to reason.

On Understanding the Numbers

People who can't understand numbers are useless. The gemba where numbers are not visible is also bad. However, people who only look at the numbers are the worst of all.

On Costs

Costs do not exist to be calculated. Costs exist to be reduced.

On Work Worthy of Humans

I think it ruins people when there is no race to get each person to add their good ideas to the work they do within a company. Your improvements make the job easier for you, and this gives you time to make further improvements. Unlike in the [Charlie] Chaplin movie where people are treated as parts of a machine, the ability to add your creative ideas and changes to your own work is what makes it possible to do work that is worthy of humans.

On Kaizen

Kaizen ideas are infinite. Don't think you have made things better than before and be at ease. As I mentioned earlier, this would be like the student who becomes proud because they bested their master two times out of three in fencing. Once you learn how to pick up the sprouts of kaizen ideas it is important to have the attitude in our daily work that just underneath one kaizen idea is yet another one.

On Patience

I was young and very eager but I saw that pushing sudden changes over a short period of time was not a good plan so I decided to stay calm and proceed deliberately.

On Taking His Advice

You are a fool if you do just as I say. You are a greater fool if you don't do as I say. You should think for yourself and come up with better ideas than mine.

References

Toyota Production System: Beyond Large-Scale Production, Taiichi Ohno, Productivity Press, 1988.

Taiichi Ohno's Workplace Management, Taiichi Ohno, McGraw-Hill, 2013.

Inside the Mind of Toyota: Management Principles for Enduring Growth, Satoshi Hino, Productivity Press, 2005.

The Birth of Lean, Koichi Shimokawa and Takahiro Fujimoto, Lean Enterprise Institute, Inc., 2009.

gijutsu saikoushoku ga kataru toyota wa kou hito wo sodateru, 技術最高職が語る「トヨタはこう人を育てる」, Nikkei Business Online, August 4, 2008.

toyota tsuyosa no genten ohno taiichi no kaizen damashii, トヨタ強さの原点　大野耐一の改善魂, Nikkan Kogyo Shimbunsha, 2005.

zubari gemba no muda dori jiten – toyota seisan houshiki no jissen tetsugatku, ズバリ現場のムダどり事典　―　トヨタ生産方式の実践哲学, Hitoshi Yamada, Nikkan Kogyo Shimbunsha, 1989.

Original research by Jon Miller of Japanese source materials from Toyota and recollections of veterans from Toyota and group companies, 1993–2012.

A NOTE ON TRANSLATION
FROM JAPANESE TO ENGLISH

Our philosophy was to translate both Ohno's meaning and style, and sacrifice neither of these to polish the English expression. Japanese is a language that relies heavily on context, and can often be vague. In the process of translation and editing, our goal was to cut out nothing, and add as little as possible in order to maintain the flow of Ohno's speech and thought during this interview.

I hope the reader will not be put off by the many Japanese words in the text. Ohno calls for a "revolution of awareness" and makes his point by telling stories about the misconceptions people have. Often these stories are illustrated by pairs of ideas, such as "reduced volume" versus "limited volume" or "automatic" versus "autonomous," which mean different things but are spoken the same way or written nearly the same way in Japanese. Ohno uses these pairings to help us think differently about work and how we manage it. His expressions may be uniquely Japanese, but these ideas are universal.

Where the author introduces terms or concepts without explaining them in detail, we have added footnotes throughout the text. However, we did not provide full and complete explanations of these concepts. This was not the author's intent, nor was this within the scope of this translation. These footnotes are not present in the original text. They are intended to provide additional context or detail where it is implied or required by the text, but not to change the meaning of the text in any way.

In most cases where the author uses the Japanese word "*gemba*," it has been left as *gemba*. This concept is very important, and the reader can better appreciate Ohno's ideas by taking the word *gemba* and filling in the meaning, which in context can mean "shop floor," "workplace," or "the people on the shop floor." *Gemba* is important, and this book, whose Japanese title is "Taiichi Ohno's Management of the *Gemba*," offers a unique perspective into the author's philosophy.

The reader will find both *Toyota* as well as *Toyoda* in this text. The family name Toyoda (with the letter "d") is still used today, and several names of Toyota group companies bear the Toyoda name. The name of the automobile company Toyota was changed from Toyoda in 1936, because the number of strokes (eight) to write Toyota was a luckier number than writing Toyoda (ten strokes), because the founder Kiichiro Toyoda wanted to make a distinction between his public and private life, and because "Toyota" has a cleaner sound.

Jon Miller

INDEX

ABOUT KAIZEN INSTITUTE

Founded by Masaaki Imai in 1985, Kaizen Institute is the pioneer and global leader in promoting the spirit and practice of *kaizen*. Its global team of professionals is dedicated to building a world where it is possible for everyone, everywhere, every day, to be able to "*kaizen* it."

Kaizen Institute guides organizations to achieve higher levels of performance in the global marketplace—easier, faster, better, and cheaper. Kaizen Institute *sensei* challenge clients to help develop leaders capable of sustaining continuous improvement in all aspects of their enterprise. Kaizen Institute creates a worldwide community of practice in *kaizen*.

The major services of Kaizen Institute include:

Consulting and Implementation

▲ Partnering with clients for long-term *kaizen* implementation
▲ Operating system design and deployment
▲ Breakthrough projects and turnarounds

Education and Training

▲ Business training, academic, and online training curriculum design
▲ *Kaizen* practitioner, coach, and manager level certification
▲ On-site training, workshops, and seminars

Tours and Benchmarking

▲ "Kaikaku" benchmark to best-in-class organizations in Japan and worldwide
▲ Building peer-to-peer learning and tour exchange network

Visit www.kaizen.com to learn more about *kaizen* and the world-changing purpose of Kaizen Institute.

WORLDWIDE CONTACT INFORMATION FOR KAIZEN INSTITUTE CONSULTING GROUP

Americas

Kaizen Institute USA

7137 East Rancho Vista Drive,
  B-11
Scottsdale, AZ 85251 USA
Tel: +1 480 320 3476
Fax: +1 480 320 3479
Email: usa@kaizen.com

Kaizen Institute México

Av. Chapultepec 408
Int. 3 Colinas del Parque
78260 México
Tel: +52 444 1518585
Email: mx@kaizen.com

Kaizen Institute Brazil

Al. dos Jurupis
452 – Torre A – 2º.
Andar 04088-001
São Paulo – SP Brazil
Tel: +55 (11) 5052 6681
Fax: +55 (11) 5052 6681
Email: br@kaizen.com

Kaizen Institute Chile

Av. Providencia 1998 of. 203
Providencia, Santiago, Chile
Tel: +52 (0) 2 231 1450
Email: cl@kaizen.com

Asia Pacific

Kaizen Institute Japan

Glenpark Hanzomon, #310
2-12-1 Kojimachi
Chiyoda-ku, Tokyo 102-0083
Japan
Tel: +81 (0) 3 6909 8320
Fax: +81 (0) 3 6909 8321
Email: jp@kaizen.com

Kaizen Institute China

1027 Chang Ning Road,
  Suite 2206
Shanghai, China
Tel: +86 (0) 21 6248 2365
Email: cn@kaizen.com

Kaizen Institute Singapore

20 Cecil Street
#14-01 Equity Plaza
Singapore 049705
Tel: +65 (0) 6305 2410
Email: sg@kaizen.com

Kaizen Institute New Zealand

15a Vestey Drive,
  Mt Wellington
Auckland 1060 New Zealand
Tel: +64 (09) 588 5184
Email: nz@kaizen.com

Kaizen Institute India

Office No. 1A, Second Floor
Sunshree Woods Commercial
  Complex, NIBM Road
Kondhwa 411 048 Pune, India
Tel: +91 92255 27911
Email: in@kaizen.com

Europe, Middle East, and Africa

Kaizen Institute Germany

Werner-Reimers-Strasse 2-4
D-61352 Bad Homburg
Germany
Tel: +49 (0) 6172 888 55 0
Fax: +49 (0) 6172 888 55 55
Email: de@kaizen.com

Kaizen Institute France

Techn'Hom 3
15 Rue Sophie Germain
F-90000 Belfort France
Tel: +33 145356644
Fax: +33 145356564
Email: fr@kaizen.com

Kaizen Institute United Kingdom

Regus House
Herald Way
Pegasus Business Park
Castle Donington DE74 2TZ
UK
Tel: +44 (0) 1332 6381 14
Email: uk@kaizen.com

Kaizen Institute Netherlands

Bruistensingel 208
5232 AD
's-Hertogenbosch Netherlands
Tel: +31 (0)73 700 3440
Email: nl@kaizen.com

Kaizen Institute Spain

Ribera del Loira, 46 Edificio 2
28042 Madrid, Spain
Tel: +34 91 503 00 19
Fax: +34 91 503 00 99
Email: es@kaizen.com

Kaizen Institute Switzerland

Bahnhofplatz
Zug 6300 Switzerland
Tel: +41 (0) 41 725 42 80
Fax: +41 (0) 41 725 42 89
Email: ch@kaizen.com

Kaizen Institute Portugal

Rua Manuel Alves Moreira, 207
4405-520 V.N.Gaia Portugal
Tel: +351 22 372 2886
Fax: +351 22 372 2887
Email: pt@kaizen.com

Kaizen Institute Italy

Piazza dell'Unità, 12 40128
Bologna, Italy
Tel +39 051 587 67 44
Fax: +39 051 587 67 73
Email: italy@kaizen-institute.it

Kaizen Institute Kenya

c/o KAM – Kenya Association
   of Manufacturers
3 Mwanzi Road
Opp Nakumatt Westgate
   Westlands
Nairobi, Kenya
Tel: +254722201368
Email: afr@kaizen.com

Additional Kaizen Institute Locations

Austria, Belgium, Czech
   Republic, Hungary,
   Malaysia, Poland, Romania,
   Russia

To find contact information
for all locations, please visit
www.kaizen.com.

Gemba Academy Online Training

www.GembaAcademy.com

Kaizen Institute Blog

www.gembapantarei.com

Executive Master's Degree Program

www.kaizen.com/Master